Special Structural Topics

Special Structural Topics covers specialty structural situations for students and professional architects and engineers, such as soil mechanics, structural retrofit, structural integrity, cladding design, blast considerations, vibration, and structural sustainability. As part of the Architect's Guidebooks to Structures series, it provides a comprehensive overview, using both imperial and metric units of measurement and with more than 150 images. As a compact summary of key ideas, it is ideal for anyone needing a quick guide to specialty structural considerations.

Paul W. McMullin is an educator, structural engineer, and photographer. He holds degrees in mechanical and civil engineering and is a licensed engineer in numerous states. He is a founding partner of Ingenium Design, providing innovative solutions to industrial facilities. Currently an adjunct professor at the University of Utah in Salt Lake City, USA, he has taught for a decade and loves bringing project-based learning to the classroom.

Jonathan S. Price is a structural engineer and adjunct professor at Philadelphia University in Pennsylvania, USA, where he was honored with the Distinguished Adjunct Faculty Award in 2006. He holds a Bachelor of Architectural Engineering degree from the University of Colorado, USA, a Master of Science degree in civil engineering from Drexel University in Philadelphia, USA, and is registered in 12 states.

Sarah Simchuk is a project architect and fine artist working towards architectural licensure in large-scale retail design. She holds a Bachelor's and Master's degree in Architecture from the University of Utah. She is in the early stages of her architectural career, with an inclination towards design and details in project management. She comes from a fine art background, with more than 15 years' experience in hand drawing and rendering, and lends a 3-D approach to the understanding of structures.

Architect's Guidebooks to Structures

The *Architect's Guidebooks to Structures* series addresses key concepts in structures to help you understand and incorporate structural elements into your work. The series covers a wide range of principles, beginning with a detailed overview of structural systems, material selection, and processes in *Introduction to Structures*, following with topics such as *Concrete Design*, *Special Structural Topics*, *Masonry Design*, and *Timber Design*, and finishing with *Steel Design*, to equip you with the basics to design key elements with these materials and present you with information on geotechnical considerations, retrofit, blast, cladding design, vibration, and sustainability.

Designed as quick reference materials, the *Architect's Guidebooks to Structures* titles will provide architecture students and professionals with the key knowledge necessary to understand and design structures. Each book includes imperial and metric units, rules of thumb, clear design examples, worked problems, discussions on the practical aspects of designs, and preliminary member selection tables, all in a handy, portable size.

Read more in the series blog: http://architectsguidestructures.wordpress.com/

Introduction to Structures
Paul W. McMullin and Jonathan S. Price

Concrete Design
Paul W. McMullin, Jonathan S. Price, and Esra Hasanbas Persellin

Special Structural Topics
Paul W. McMullin, Jonathan S. Price, and Sarah Simchuk

Masonry Design
Paul W. McMullin and Jonathan S. Price

Timber Design
Paul W. McMullin and Jonathan S. Price

Steel Design
Paul W. McMullin, Jonathan S. Price, and Richard T. Seelos

Special Structural Topics

Edited by
Paul W. McMullin,
Jonathan S. Price,
and Sarah Simchuk

Routledge
Taylor & Francis Group
NEW YORK AND LONDON

First published 2018
by Routledge
711 Third Avenue, New York, NY 10017

and by Routledge
2 Park Square, Milton Park, Abingdon, Oxon OX14 4RN

Routledge is an imprint of the Taylor & Francis Group, an informa business

Library of Congress Cataloguing in Publication Data
Names: McMullin, Paul W., editor. | Price, Jonathan S., editor. | Simchuk, Sarah, editor.
Title: Special structural topics / [edited by] Paul W. McMullin, Jonathan S. Price, and Sarah Simchuk.
Description: New York, NY : Routledge, 2018. | Series: Architect's guidebook to structures | Includes bibliographical references and index.
Identifiers: LCCN 2016056748| ISBN 9781138838901 (hardback) | ISBN 9781138838925 (pbk.) | ISBN 9781315733722 (ebook)
Subjects: LCSH: Structural engineering.
Classification: LCC TA645 .S57 2018 | DDC 624.1—dc23
LC record available at https://lccn.loc.gov/2016056748

ISBN: 978-1-138-83890-1 (hbk)
ISBN: 978-1-138-83892-5 (pbk)
ISBN: 978-1-315-73372-2 (ebk)

Typeset in Calvert
by Florence Production Ltd, Stoodleigh, Devon, UK

For our partners
Kay
Kari
Kristoff

Contents

Acknowledgments

Like previous books in the series, this volume wouldn't be what it is without the diligent contributions of many people. We thank Hannah Vaughn, and Teran Mitchell for their wonderful figures and diligent efforts. We are very grateful to each chapter contributor for sharing the depth of their knowledge, and to Steve Judd for his review of the cladding chapter.

We thank Wendy Fuller, our commissioning editor, Norah Hatch, our editorial assistant, Louise Smith, our copy editor, and Alanna Donaldson, our production editor. Each of you has been wonderful to work with and encouraging and helpful along the journey. We thank everyone at Routledge who produced and marketed the book. We are also very grateful to the production team at Florence Production, particularly Laurence Paul and Ellie Jarvis.

A special thanks to our families, and those who rely on us, for being patient when we weren't around.

We are unable to fully express our gratitude to each person involved in preparing this book. It is orders of magnitude better than it would have otherwise been, thanks to their contributions.

Contributors

EDITORS

Paul W. McMullin is an educator, structural engineer, and photographer. He holds degrees in mechanical and civil engineering and is a licensed engineer in numerous states. He is a founding partner of Ingenium Design, providing innovative solutions to industrial facilities. Currently an adjunct professor at the University of Utah in Salt Lake City, USA, he has taught for a decade and loves bringing project-based learning to the classroom.

Jonathan S. Price is a structural engineer and adjunct professor at Philadelphia University in Pennsylvania, USA, where he was honored with the Distinguished Adjunct Faculty Award in 2006. He holds a Bachelor of Architectural Engineering degree from the University of Colorado, USA, a Master of Science degree in civil engineering from Drexel University in Philadelphia, USA, and is registered in 12 states.

Sarah Simchuk is a project architect and fine artist working towards architectural licensure in large-scale retail design. She holds a Bachelor's and Master's degree in Architecture from the University of Utah. She is in the early stages of her architectural career, with an inclination towards design and details in project management. She comes from a fine art background, with more than 15 years' experience in hand drawing and rendering, and lends a 3-D approach to the understanding of structures.

CONTRIBUTORS

Tracy Rae-Baker Aragon, P.E., is a civil engineer in the U.S. Army Corps of Engineers. As a former member of the Geotechnical Engineering Section, she has experience designing and constructing airfield pavements, slope stabilization projects, and foundations. She has also worked in dam and levee safety, inspecting and assessing flood damage reduction projects.

She currently works in the Dam Operations group, maintaining and operating large Corps-owned dams, one of which she comfortably lives downstream of.

Caroline Field, M.Sc., B.Eng. (Hons), C.Eng. M.I.C.E., P.E., Member RSES, leads BuroHappold's Risk and Resilience Team. Caroline has more than 20 years of experience designing structures for the built environment, including the Millennium Dome (O2 Arena) in London, the California Academy of Sciences in San Francisco, and the National Geospatial Agency Headquarters in Virginia. She spent the past 17 years specializing in providing protective, resilient design for a variety of buildings and structures to resist the effects of earthquake, blast, and progressive collapse. She spent 12 years in the United States working closely with the federal government on projects that have mandated blast requirements and returned to the UK in August 2013. Prior to rejoining BuroHappold, Caroline worked as a member of the senior management team at Hinman Consulting Engineers in San Francisco for 6 years and at Arup for the previous 6 years.

Rob Fleming, AIA, LEED AP BD+C, is a registered architect who lives in Philadelphia. His 25-year mission is to seek out new ways that sustainable design can transform our lives and to boldly go where everyone needs to go—towards a sustainable future. Rob is the co-creator and director of the award-winning Master of Science in Sustainable Design program at Philadelphia University. He is also a Sustainability Fellow at Re:Vision Architecture in Philadelphia, where he consults and facilitates integrated design charrettes.

Justin W. Jacobs, AIA, LEED AP, is a principal and senior project manager at GSBS Architects in Salt Lake City, UT. With more than 25 years of experience in the field of architecture, he has worked on public- and private-sector projects, in virtually every capacity from project inception to completion. Justin is a licensed architect in the State of Idaho and member of the A.I.A. and currently serves on the Utah Board of the Design Build Institute of America.

Kevin J. LaMalva, P.E., is a structural engineer and fire protection engineer. He holds degrees in civil and fire protection engineering and is a licensed civil and fire protection engineer. He is chair of the ASCE/SEI Fire Protection Committee and is a member of several other industry committees that pertain to structural fire engineering. He is the co-author of ASCE/SEI 7–16 Appendix E and editor of the upcoming ASCE/SEI Guideline: Structural Fire Engineering. Currently a consultant, he has worked on a multitude of design and investigative projects in both structural and fire protection engineering, with a long-standing passion for when these disciplines converge.

Teran Mitchell received his Bachelor's degree in Architectural History from Virginia Commonwealth University and a Master's degree in Architecture from the University of Utah. He is an active member of the A.I.A., NCARB, and Society of Architectural Historians. His studies and travels have led him throughout North and South America, where he gained an appreciation for the power of culture to influence architecture, and the beauty derived from regional and cultural architectural variation. Teran and his wife, Michelle, are both natives of northern Utah. They have five children and currently live in Richmond, Virginia, where Teran is Vice President at Ballou Justice Upton Architects.

Luke Pascoe, M.Eng. (Hons.) ACGI C.Eng. M.I.C.E., Member of the RSES, is a civil and structural engineer specializing in the design and assessment of structures subjected to blast loads. He studied at Imperial College London and at the École Nationale des Ponts et Chaussées in Paris. After several years as a general structural engineer, he became increasingly involved in assessments of structural response to blast loads, a field he has been continuously involved in for the past 5 years. He has extensive experience in the design of blast-resistant glazing systems and has coordinated research and testing with Imperial College London and the European Commission and published several peer-reviewed articles on the subject.

Hannah Vaughn is an architect based in Salt Lake City, UT. She is a dedicated practitioner and teaches as adjunct faculty at the University of Utah School of Architecture.

Esra Hasanbas Persellin moved to the US from Turkey, following her passion for engineering. She holds an M.S. degree from the University of Utah. Her studies specialize in vibration analysis and control. She has published articles and has presented at several conferences. She is a structural engineer at Bentley Systems, supporting analytical modeling (FEA) products.

Saglinda H. Roberts, ASID, CID, LEED AP BD+C, is an educator and designer who is deeply convinced of architecture's ability to transform lives on all levels. Her focus is creating a deeper, more holistic understanding of the dynamic goals of aesthetics, culture, performance, and the ecological aspects of each site to promote regenerative, authentic sustainable design. Saglinda is an Assistant Professor at Kean University, where she teaches sustainable design and the upper design studios.

Introduction

As buildings become more complex, our need to understand a wider variety of specialty structure topics increases. A century ago, architectural and structural drawings were one and the same. Today, on highly complex structures, we may see drawings from several structural firms, and specialty design input from a number of others. This specialty engineering may include vibration, blast, seismic, retrofit, structural integrity, cladding, and fire. Additionally, sustainability, soil mechanics, and quality control are of fundamental importance to successful buildings and are equally complex.

This book is different than the others in the series. It provides a high-level overview of each specialty topic to give the practicing architect a working knowledge of possible design elements on their structure. This will provide them enough technical know-how to coordinate the work. This text also provides the engineer who may be new to these topics a sound starting point from which to build their understanding.

This book is authored by practicing engineers and architects, deeply experienced in their area of expertise. They understand the technical nuances and the big picture of how a specialty project goes together. Their desire is to bridge the gap between structural engineering and architecture—a gap that historically didn't exist and is unnecessarily wide today. We hope it opens the door for you.

Vibration

Chapter 1

Esra Hasanbas Persellin

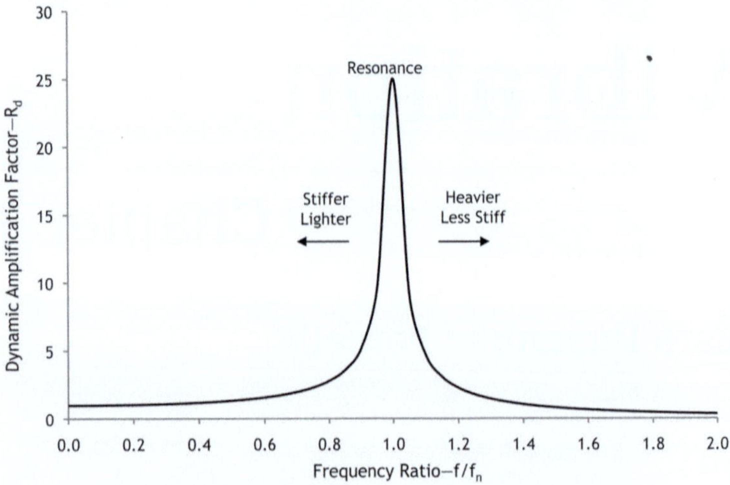

Figure 1.1 Dynamic Amplification Factor

Vibration problems can be vexing at best, and life-threatening at worst. They can seem like a serious mystery if we don't have the right analytical and experimental tools. With a grasp of basic **dynamic** principles, we can understand the office floor that bounces every time someone walks by, or the large rotating machine that seems to be bringing down the house.

Unlike strength or deflection design, where we want to be under the maximum permitted load (away from failure), safe vibration design is on either side of **resonance**, illustrated in Figure 1.1. This is where much of the challenge comes from. A change in **mass** or **stiffness** may shift us towards or away from resonance, depending on where we started. Vibration theory allows us to understand how variation of the input changes the structural response.

1.1 THEORY

1.1.1 Load Types

You are familiar with static **loads**: dead, live, or snow. They vary slowly with time. Dynamic loads, on the other hand, change rapidly and are repetitive or **transient** in nature. We divide dynamic loads into periodic and non-periodic loads. Periodic loads are **harmonic**, often **sinusoidal**, loads. A rotating machine or a rhythmic human activity such as aerobics is an example of a harmonic load. Non-periodic loads may be impact loads or bomb blasts.

In this chapter, we will be focusing on harmonic loads caused by rotating machinery and human activity. Let's begin with a discussion of key dynamics concepts.

1.1.2 Mode Shapes

Structures, such as strings, vibrate at different **frequencies** based on material and geometric stiffness. Each fundamental frequency of vibration is known as a **mode**. If we take the string in Figure 1.2, we see it vibrates in a parabolic shape from end to end (mode 1), from end to midpoint (mode 2), and in thirds (mode 3).

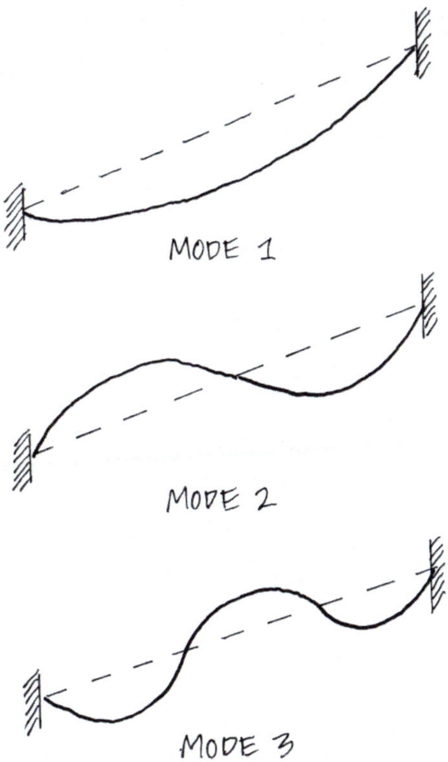

Figure 1.2 Mode Shapes of a String

Each mode shows us a different vibration **shape**. Figure 1.3 shows a mode shape associated with one of the fundamental vertical natural frequencies for a simple steel structure, in which we see hints of the vibrating string. Diagrams of the deflected shape help us visualize what is happening when we are designing for vibration.

Generally, we are concerned about the lower modes; however, depending on structures' mass and stiffness distribution, the higher modes can become important as well.[1]

1.1.3 Frequency

Frequency is a measure of vibration rate. We measure it in Hertz (Hz, cycles per second) or revolutions per minute (rpm). We concern ourselves with forcing and **natural frequencies**. Equipment rotation speed and human activities are examples of forcing (exciting) frequency—they drive the problem. Natural frequency represents the rate at which a structure vibrates if there are no continuous external forces present on a structure[2]—they determine response to the input.

Mathematically, natural frequency is defined as:

$$f_n = \frac{1}{2\pi}\sqrt{\frac{k}{m}} \tag{1.1}$$

where:

f_n = natural frequency of the structure, Hz
k = stiffness of the structure, lb/in (N/m)
m = mass of the structure, lb (kg)

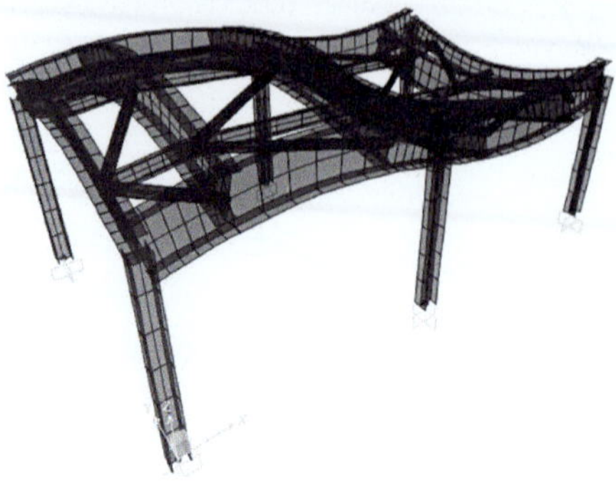

Figure 1.3 A Mode Shape of a Simple Steel Structure

Or, it can be defined as:

$$f_n = \frac{1}{2\pi}\sqrt{\frac{g}{\delta}} \tag{1.2}$$

where:

g = acceleration of gravity, 386.4 in/s^2 (9.81 m/s^2)

δ = static deflection, in (mm)

Natural frequency is a function of stiffness and mass. The more material, the greater the stiffness, the greater the frequency. The more mass, the lower the frequency.

Equation (1.2) will be more familiar to structural engineers, who are accustomed to calculating deflections. It also captures the compound effects of multiple deflection sources, such as girders and columns.

Figure 1.4 shows the frequency solutions for beams with different end conditions. These simple frequency solutions provide a starting point for vibration analysis, or for checking complex software solutions.

1.1.4 Resonance

When the forcing and natural frequencies get close to each other, resonance occurs. The result of being close to resonance may only be annoying, such as a floor vibration, but it also may cause a machine to malfunction, or lead to complete structural collapse.

The Tacoma Narrows Bridge is a good example of the forcing and natural frequencies getting too close. Tacoma Narrows Bridge was a suspension bridge with a total length of 5,939 ft (1,810 m). Called Galloping Gertie, the bridge vibrated vertically when the wind was blowing around 42 mph (67.6 km/h), as shown in Figure 1.5. The engineers did not consider the **aerodynamic** forces and designed a bridge whose natural frequency in both the vertical and torsional directions was too close to the wind-**induced frequency**. Four months after it opened, the bridge went into resonance and collapsed, on November 7, 1940.

1.1.5 Damping

Vibration in a structure eventually diminishes if there is no continuous input. This is owing to **damping**—that is, energy-dissipating mechanisms. Damping can include friction between structural and nonstructural

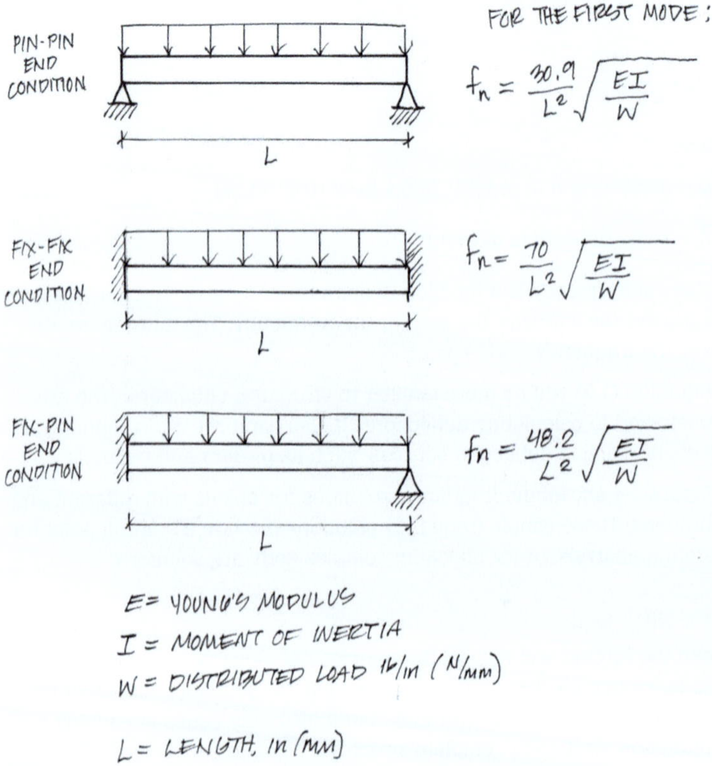

Figure 1.4 Natural Frequency Equations of a Beam with Different End Conditions

members, friction at the connections, micro-cracks in concrete, or viscous dampers such as those in a vehicle.

We use the damping ratio ξ to analytically account for the effect of damping on the structure's response. It is 1–2 percent for steel frame structures, 4–7 percent for reinforced concrete structures, and 4–12 percent for typical office buildings.[3]

Theoretically, a structure with no damping will vibrate endlessly, as shown by the graph in Figure 1.6. However, real structures have inherent damping. Figure 1.7 shows how the vibration dissipates for a structure with a 3 percent damping ratio.

Figure 1.5 Tacoma Narrows Bridge Twisting before Collapse

Source: Courtesy University of Washington Libraries, Special Collections, UW21413

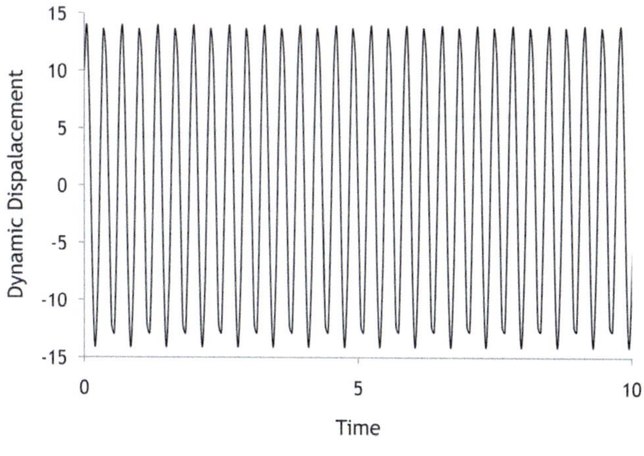

Figure 1.6 Acceleration–Time History without Damping

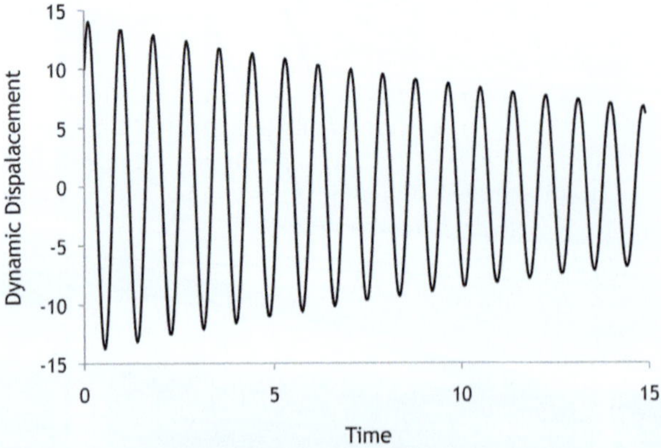

Figure 1.7 Acceleration–Time History with Damping

1.1.6 Deflection Amplification

When a cyclical dynamic force acts on the structure, the static displacement is amplified by a given amount, depending on how close the loading frequency is to resonance. This is the **dynamic amplification factor**, R_d, plotted in Figure 1.1. It is a function of damping, natural frequency, and the **forcing frequency** from machinery or human activity. We calculate it as:

$$R_d = \frac{1}{\sqrt{\left[\left|1-\left(f/f_n\right)\right|^2\right]^2 + \left[\left|2\xi\left(f/f_n\right)\right|\right]^2}} \tag{1.3}$$

where:

R_d = dynamic amplification factor
f_n = natural frequency of the structure, Hz
f = induced frequency, Hz (operating machinery, human activity)
ξ = damping ratio

To find the dynamic deflection, we multiply the static displacement by the dynamic amplification factor, as follows:

$$u_o = R_d u_{st} \tag{1.4}$$

where:

u_o = dynamic displacement
u_{st} = static displacement

1.2 VIBRATION CAUSES

Sometimes, one machine will operate in one structure without causing any problems, and yet, in another, will shake the members virtually out of control. There are several reasons behind this seeming mystery. First is the ratio of the operating machine frequency to the structure's natural frequency in the direction of operation. Recalling the amplification factor (see Equation 1.3), we notice that the frequency ratio plays a big role in dynamic displacement calculations. As the frequency ratio gets closer to 1, the amplification factor peaks, as shown in Figure 1.1.

Second, low structural stiffness causes higher static deflection. If the structure has a large static deflection, then a small dynamic amplification can cause excessive vibration, as borne out by Equation (1.4).

Keep in mind that changes in weight and stiffness will have seemingly contradictory effects, depending on which side of the curve in Figure 1.1 we find ourselves. If we are on the right, and we reduce mass, we will get closer to resonance. Conversely, if on the left, and we decrease stiffness or add mass, we may also get closer to resonance.

For example, the author worked on a project where the power plant owner replaced a crusher with a much lighter one and removed a 30,000-lb (13,608-kg) weight from the top of the structure. Everyone involved thought it made sense to reduce the weight, and the structure would be happier. However, as soon as they started the crusher, the entire structure started swaying. The reduction of mass and removal of a bottom-story brace combined to make an entire structure vibrate that had previously been steady.

Even though the vibration due to human activity shares the same theory as vibration due to machinery, their design priorities are different. Structural integrity and mechanical efficiency are priorities for the control of machine-induced vibration, whereas human-induced vibration prioritizes discomfort and the integrity of sensitive equipment. Therefore, the acceptable limits for these two types of vibration vary.

1.2.1 Equipment-Induced Vibration

Rotating machinery generates harmonic loading on its support structure. Depending on the frequency ratio and the static displacement, this might lead to excessive vibration in the structural support frame and could result in failure of the structural system.

Rotating machinery commonly causes vertical vibration problems in structures, as their operating frequency range falls within the natural frequency of common structures. However, this doesn't mean that we should only investigate the vertical response of a structure. We need to consider the potential of horizontal vibration by paying attention to the machine frequency and the weight and stiffness of the structure.[4] We will use all these parameters in Section 1.3, where we discuss how to design for vibration.

Table 1.1 shows the frequencies of common equipment types to demonstrate the frequency range that we are dealing with. For example, many steel support frames have similar frequency ranges to low-speed screens. Not a helpful combination.

1.2.2 Human Activity Vibration

Human activities such as walking, dancing or aerobics can cause vibration in a structure. For an engineer, the vibration effects of a person walking on your slab is low on your list of concerns when you are dealing with seismic, wind, or snow loads. However, we still need to examine possible vibration serviceability issues.

Vibration acceptance benchmarks for human activity vary, depending on the structure's purpose. For example, when we design a floor in a hospital with sensitive equipment or a microscopy lab, the floor vibration limits are much tighter (measured in micro-inches per second) than a for gym floor. In an office setting, the perception of vibration will vary from person to person. The American Institute of Steel Construction (AISC) has published vibration acceptance criteria for different cases. These are summarized in Figure 1.8 for outdoor bridges, malls, and living areas.

Table 1.1 Representative Forcing Frequencies for Vibratory Equipment

Equipment	*f (Hz)*
Screens (high speed)	500–5,000
Screens (low speed)	2–12
Compactors	< 1
Turbines	40–300
Gear boxes	0–3
Crushers	2–15

Source: Scott Adams

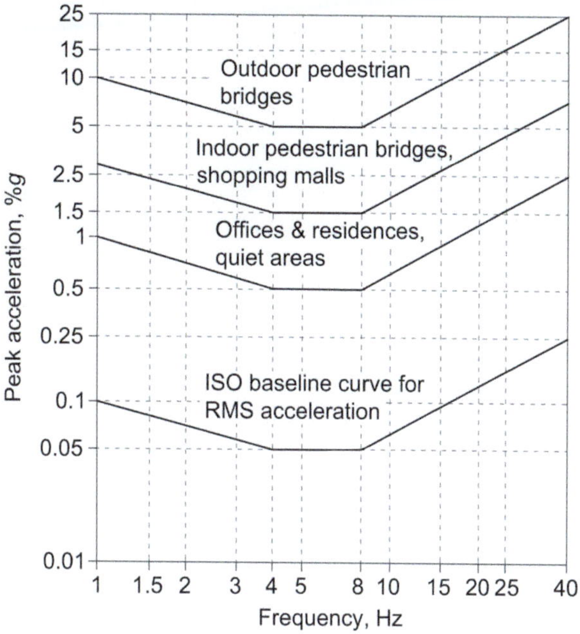

Figure 1.8 Peak Acceleration Limits for Human-Induced Vibration

Source: Copyright © American Institute of Steel Construction. Reprinted with permission.

Walking at a certain pace exerts a periodic force on a structure. This force generally generates a low frequency of about 1.6–2.2 Hz. However, periodic steps also generate harmonic waves, one after another. This can multiply the step frequency and make the structure susceptible to higher modes or frequencies.

Aerobic activities exert higher-frequency loading with greater amplitudes. However, owing to the active nature of the activity, humans are less sensitive to the floor vibrations, and so the acceptable acceleration limits are higher. Dancing, jumping, or aerobics cause a loading frequency of 2–8.3 Hz.

Table 1.2 shows the possible forcing frequencies generated owing to walking and rhythmic excitations. One may use these values and compare them with the floor's natural frequency to avoid resonance. Also, one can compare the peak accelerations with recommended limits (Figure 1.8) to avoid dynamic amplitude. However, even before calculations are done, primary walkways should be located away from sensitive areas or equipment. One study[5] shows that breaking the length of corridors actually decreases vibration amplitudes.

The duration of vibration exposure is also a key element in vibration design. If the floor is out of the recommended range, think about how long the duration of vibration is. If the duration is quite short, the floor acceleration could be acceptable.

1.3 VIBRATION DESIGN

Now we know all the scary structural dynamics terms, and one might think this is where we dive into differential equations. Luckily, this is not the case. In this section, we tackle vibration design using frequency analysis steps.

1.3.1 Design Steps

Considering the potential for excessive vibration in your structure at the design stage is crucial. Once the structure is built, retrofits are expensive and time-consuming. The following steps will guide you in designing a reliable structure and avoiding excessive vibration problems:

1. Calculate the natural frequency of your structure. For a simple analysis, use the equations in Figure 1.4. For more complex structures,

Table 1.2 **Forcing Frequencies for Human-Induced Vibration**

Input	*i*	*f, Hz*
Walking	1	1.6–2.2
	2	3.2–4.4
	3	4.8–6.6
Aerobic class	1	2–2.8
	2	4–5.5
	3	6.8–8.3
Group dancing	1	1.5–3

Notes: i = harmonic multiple; α_i = dynamic coefficient

use structural analysis software. Only include the minimum realistic dead loads in your model. This will include the weight of the equipment, permanent office furniture, etc.

2. If you are designing for a rotating machinery support, you will need its operating frequency and its static weight.
3. For vertical vibration cases, model one or two bays of the structure where the machinery is operating. For horizontal cases, model the entire structure.
4. Run your model. Review the mode shapes and each natural frequency.
5. From your model, obtain the natural frequencies that are in the direction and the range of the machinery's operating frequency. Be careful about comparing apples with apples—that is, do not compare the structure's vertical natural frequency with the machine's horizontal operating frequency.
6. Calculate the frequency ratio (f/f_n) and corresponding amplification actor, R_d.
7. Calculate the dynamic response by multiplying the largest static displacement by the amplification factor.
8. If the frequency ratio is close to 1, make changes to the structural system to either increase or decrease its natural frequency and avoid resonance. Recall that the structure's natural frequency is a function of its mass and its stiffness. Figure 1.9 shows how changing a structure's

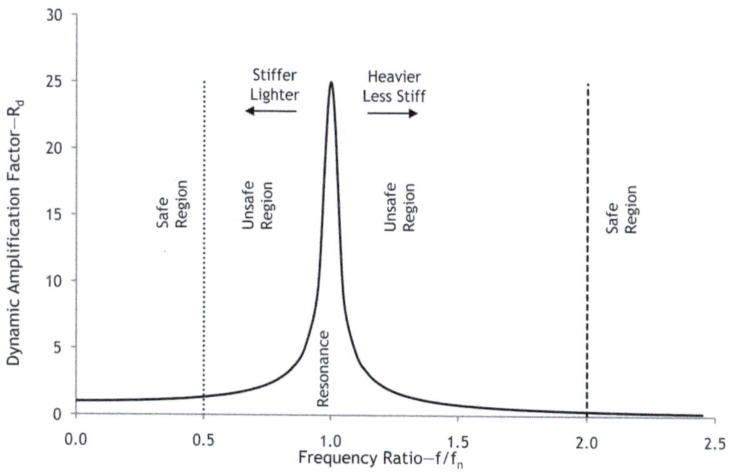

Figure 1.9 Frequency Ratio Acceptance Criteria

mass or stiffness affects the dynamic amplification. Depending on which side of the peak your system is on, you can pick the most suitable solution. For example, if you are just to the left side of the peak, you can always increase your beam sizes, or use more bracing to increase the stiffness. However, if your system is on the right side of the peak, you can increase mass to get further away from resonance. However, keep in mind that static displacement plays a big role in vibration. Increasing mass, or decreasing stiffness, may shift the frequency ratio farther to the right, away from the resonance, but will increase the static displacement.

9. Pay greatest attention to frequency ratio, dynamic amplification factor, and static displacement. They drive vibration design.

1.3.2 Acceptance Criteria

Once we obtain the dynamic response, we compare it with an acceptance **criterion**. For modal analysis, we use the ratio of forcing frequency to natural frequency. For a reliable design, your system should have a frequency ratio (f/f_n) smaller than 0.5 or larger than 2.0, as illustrated in Figure 1.9.

A shift of the structural system to a safe zone can be achieved by adjustment to the frequency ratio. If your system is on the left side of the peak, either increase the stiffness or decrease the mass of the structure. If on the right, increase the mass of the structure or decrease the stiffness. Pay attention to static deflection when moving to the right of resonance. Decreasing the structure's stiffness might increase the static deformation, and, even with a smaller dynamic amplification factor, the dynamic deformation will remain large.

1.3.3 Sensitivity

Whatever the analytical method, consider the sensitivity of the results to changes in input parameters (weight and stiffness). Typical stiffness calculations over-predict stiffness when compared with the actual structure. This is owing to the effect of girders, columns, and connection slip. If you are on the left side of Figure 1.9, and multiple girders and tall columns support the beam you are analyzing, the frequency ratio will shift to the right (towards resonance), as compared with your analysis. Conversely, if you are to the right of resonance and overestimated weight, your ratio will be further to the left (closer to resonance) than you anticipate.

Neither of these realities is necessarily problematic; they just require that you pay attention to your assumptions and the effect they have on the final results. Give yourself a little room on the acceptance criteria to account for variations. If you really want to know their influence, do a sensitivity study where you change the input variables according to how much they may vary in reality and see the results. This will be enlightening.

1.4 ADVANCED TECHNOLOGY

In the last decade, vibration technology has become increasingly accessible and has changed the way structural engineers handle vibration challenges. Today, **finite element analysis** (FEA) software with modal response and time-history analysis capabilities and acceleration measurements are within the reach of most moderately sized consulting firms. With the help of FEA software, we can accurately predict structural frequencies and determine the acceleration, velocity, and displacement responses. This information helps us effectively plan and design structures for vibrating loads. In addition, field measurements done by **accelerometers** provide the data necessary to determine the structure's natural frequency and its behavior under dynamic loading. Field data are invaluable for the solution of existing vibration problems and validation of analytical models.

1.4.1 Time-History Analysis

Modal analysis, discussed in the previous section, provides a simple, quick answer to vibration problems. However, it doesn't provide results as detailed as one might like. There are other analytical methods that allow us to dig deeper and obtain more detailed results.

Time-history analysis allows us to imitate the input motion, mass, and stiffness to obtain the structure's response as a function of time. This response can be deflection, velocity, or acceleration. Time-history results give us insight into what is happening each time step, rather than the overview provided by frequency ratios. For example, time-history analysis or field measurements can detect beating. This occurs when the natural frequency of a structure is close to forcing frequencies (machine's operating frequency, or rhythmic human activity), but not equal. It is defined by rapid oscillation with a slowly changing amplitude, as shown in Figure 1.10. Beating causes fatigue.

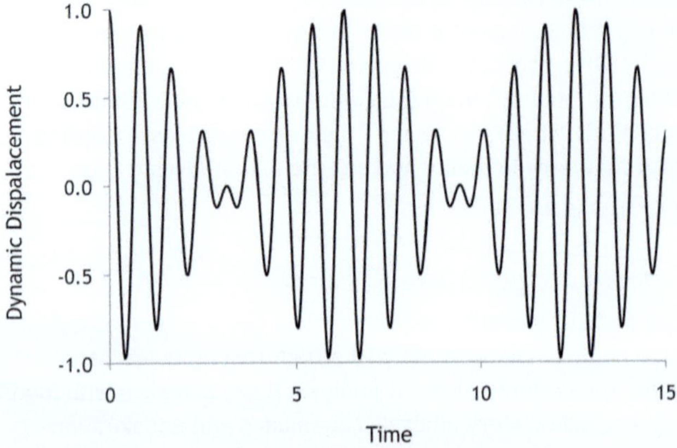

Figure 1.10 Beating Phenomena Leading to Fatigue

A number of technical organizations publish vibration limits for rotating machinery support. They are the American Petroleum Institute (API),[6] National Electrical Machinery Association (NEMA),[7] and International Standards Organization (ISO).[8] Tables 1.3 and 1.4 summarize some of these criteria. For human-induced vibration, AISC Design Guide 11[9] is a good reference, as illustrated in Figure 1.8.

Table 1.3 Acceleration Acceptance Limits for Machine-Induced Vibration

Acceleration Limits		
A < 0.05g	Excellent levels	No action required
0.05g < a < 0.10g	Good levels	No action is required unless noisy
0.10g < a <0.20g	Fair levels	No action is required unless noisy
0.20g < a < 0.40g	Rough levels	Possible action is required
0.4g < a < 1.0g	Very rough levels	Take action immediately
1.0g < a < 2.0g	Danger levels	Failure is near
a > 2.0g	Failure levels	Shut it down immediately

Table 1.4 Velocity Acceptance Limits for Machine-Induced Vibration

Velocity Limits		
$v < 0.10$ in/s	Excellent levels	No action required
$0.10 < v < 0.20$ in/s	Good levels	No action is required unless noisy
$0.20 < v < 0.35$ in/s	Fair levels	No action is required unless noisy
$0.35 < v < 0.50$ in/s	Rough levels	Possible action is required
$0.50 < v < 0.60$ in/s	Danger levels	Failure is near
$v > 0.60$ in/s	Failure levels	Shut it down immediately

1.4.2 Accelerometers

Accelerometers are effective instruments for measuring a structural dynamic response. We can instrument a structure with triaxial accelerometers and get the structure's dynamic response immediately. Triaxial accelerometers provide simultaneous, multi-axis (*x*, *y*, and *z*) measurements. Figure 1.11 shows such a sensor and the data acquisition interface used to record the measurements. Resulting acceleration data are shown in Figure 1.12, with corresponding frequencies shown in Figure 1.13.

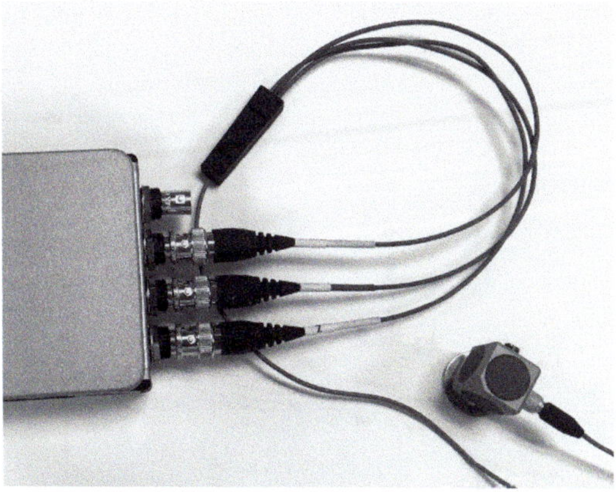

Figure 1.11 Triaxial Accelerometer and Data Acquisition System

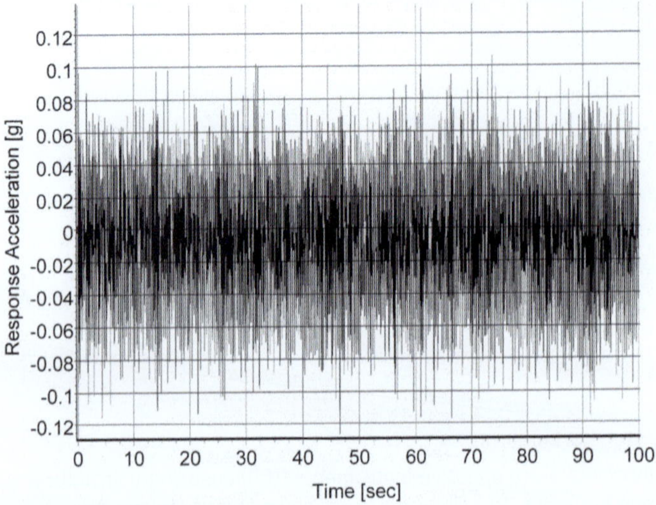

Figure 1.12 Acceleration vs. Time Readings

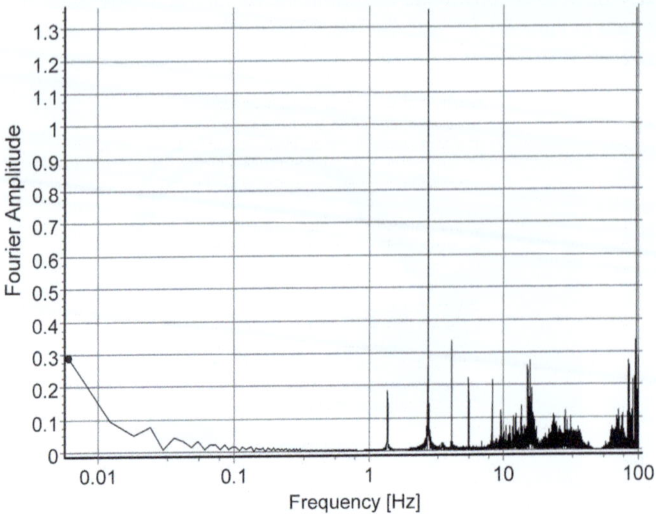

Figure 1.13 Frequency Amplitude Readings

One can obtain the natural frequency of a structure or the operating frequency of a machine by simply using this equipment. You can use this information while designing a retrofit or verifying whether the vibration is within limits or checking whether a machine is balanced.

1.5 SUMMARY

Vibration design is a topic that engineers have struggled with for many years. Hopefully, the fog is beginning to clear from the vibration mystery. You now have the basic vibration parameters, opening the door to better understanding and allowing you to work effectively towards a solution.

NOTES

1. X. Qin, N. Chouw, and T. Larkin. "Effect of Higher Modes on Structural Response with Nonlinear Soil-Foundation-Structure Interaction." Paper presented at ECCOMAS Congress, Crete, Greece, June 5–10, 2016.
2. A. Chopra. *Dynamics of Structures: Theory and Applications to Earthquake Engineering*, 3rd ed. (Upper Saddle River, NJ: Prentice-Hall, 2006).
3. AISC. *Vibrations of Steel-Framed Structural Systems due to Human Activity*, Steel Design Guide Series 11 (Chicago, IL: American Institute of Steel Construction, 2016).
4. E. Hasanbas "The Effects of Vibratory Screens on Supporting Structures." M.S. thesis, University of Utah, 2013.
5. SteelConstruction.info. "Floor Vibrations." Available at: www.steelconstruction.info/Floor_vibrations#Walking_paths (accessed May 2016).
6. API. *API Standard 670: Machinery Protection Systems* (Washington, D.C.: American Petroleum Institute, 2014).
7. ANSI. *Motors and Generators*. ANSI/NEMA MG 1–2011 (Washington, D.C.: American National Standards Institute, 2011).
8. ISO. *Mechanical Vibration of Machines with Operating Speeds from 10 to 200 rev/s*, ISO 2372 (Vernier, Switzerland: International Standards Organization, 1974).
9. AISC. *Vibrations of Steel-Framed Structural Systems Due to Human Activity*.

Blast Design

Chapter 2

Caroline Field and Luke Pascoe

Given the challenges the world faces, whether from the effects of climate change or the rise of extremism, **protective design** and **resilience** are becoming increasingly important. Explosive effects can be devastating in terms of loss of life, damage to property and reputation, and loss of business. These effects are exacerbated by poor design decisions that increase the building's **vulnerability**. For example, non-**redundant** structural systems cause failure to propagate through the structure, as was witnessed in Oklahoma City in 1995. This tragedy resulted in 168 deaths and 680 injuries. The **blast** damaged 324 buildings within a 16-block radius, destroyed or burned 86 cars, and shattered glass in 258 nearby buildings, causing estimated damage of $652 million.

Blast engineering (protective design) is not just about hardening a building; in fact, that is the last line of defense. Good blast engineering plays an integral part in the site selection, site planning, building layout, and choice of systems and materials to minimize the need for building hardening. It works with design, security, and operational measures providing **stand-off**, **access control**, and response capabilities to create a comprehensive protective strategy for the building. It is imperative that the blast consultant is engaged early in the design process. This can save the project millions of dollars through the incorporation of good practice principles into early design concepts.

This chapter introduces protective design concepts, along with advice gathered from years of experience of blast engineering. It is not intended to be a technical reference of the physics of blast or the mechanics of structures: many other publications discuss these in much greater detail. Instead, it provides valuable guidance to designers encountering blast requirements within their projects.

2.1 WHAT IS A BLAST ENGINEER?

Blast engineers focus on solutions that reduce the hazards associated with **explosion** incidents, while preserving the building's aesthetic and functional design. They assist the client to determine appropriate **risk**, performance, and design goals for the project. They model and evaluate the effects of **air-blast loading** on glazing, façade, structural elements, and people, and advise the design team (structural engineer of record (SEOR), the façade engineer, and architect) of any enhancements required to meet the defined **protection level**.

Few specific qualifications exist for blast engineers. In Europe and North America, they typically possess relevant structural professional qualifications (P.E., S.E., C.Eng., EUR Ing.). In the United Kingdom, the Register of Security Engineers & Specialists (RSES) awards professional qualifications to its members, ranging from Grade A (lowest) to Grade C (highest). It is prudent to seek an engineer with an appropriate level of experience in **blast-resistant design**.

Projects receiving U.S. federal funding may have specific blast-resistance requirements included within their contract requirements. Beyond this, it is largely at the discretion of the building owner or occupant to determine whether blast-resistance requirements should be included within a construction project. However, organizations have a duty to protect their staff and a commercial imperative to protect their assets. If a potential blast **threat** is identified, it is prudent to explore the performance of the existing design and the pros and cons of **blast mitigation measures**.

In some jurisdictions (including the UK), planning approval for some projects may depend upon the building achieving a certain level of blast performance.

2.2 THREAT AND RISK ASSESSMENTS

The first questions that a blast consultant will ask concern the explosive type and size—called **design basis threat** (**DBT**)—and the performance criteria required (level of protection). To determine if blast is a valid concern for your site, a threat and risk assessment (TARA) is conducted at the start of the project. Bear in mind that, although your building may not be a target, it may be located near a potential target.

A good TARA should identify:

- important assets and operational procedures;
- potential site vulnerabilities;
- threats that may manifest themselves over the lifetime of the asset;
- impact of these threats on people and operations;
- estimate of the overall risk environment.

The TARA should provide sufficient information to enable decisions to be made on how best to mitigate all risks across a site. It should not merely be a list of potential hazards, but should look forward over the lifetime of the building and its environment.

Potential explosive threats should be identified in the TARA, together with initial estimates of their consequences. This is used by the building

Table 2.1 Typical Range of Explosive Charge Weights

Definition	*Charge Weight TNT*
Concealed person-borne device	< 5 lb (2.5 kg) TNT
Backpack/emplaced device	< 30 lb (15 kg) TNT
Roller suitcase	< 50 lb (25 kg) TNT
Concealed vehicle-borne device	< 100 lb (50 kg) TNT
Vehicle-borne explosive device (sedan)	< 200 lb (100 kg) TNT
Vehicle-borne explosive device (SUV)	< 500 lb (250 kg) TNT
Vehicle-borne explosive device (commercial vehicle)	< 1,000 lb (500 kg) TNT
Vehicle-borne explosive device (truck)	> 1,000 lb (500 kg) TNT

owner or tenant to determine how much risk they will accept and how best to mitigate it.

When the client decides to proceed with engineered blast mitigation measures, the DBT should be chosen. These are charge sizes for which the building will be designed. They are typically of the magnitude shown in Table 2.1, and depend on the size of vehicle or package.

DBTs form one side of the blast-resistant-design equation; the other side has the defined design performance criteria. Together they form the client's blast-protection objectives.

Where no DBT is defined for the design of the building, but the client would like to provide some additional resilience, your blast consultant can provide qualitative good practice advice, such as the pros and cons of architectural and structural options, façade system types, and glazing layups and details.

Alternatively, we can back calculate the charge size that the building can withstand for a given performance criterion. This is known as an explosive consequence analysis (ECA). An example of this for a stadium façade is shown in Figure 2.1.

2.3 BLAST DESIGN CRITERIA

Once the decision has been made to protect the building from air-blast loading, the next question is: how much protection should be provided?

Damage after an explosion is almost certainly unavoidable; however, it should be proportionate to the size of the potential explosion.

Historically, the majority of fatalities not directly attributable to the explosive device (blast wave or device fragments) have come from collapse of the structure, and the majority of injuries have resulted from fragments generated from the façade and building finishes. If there is a focus on collapse prevention and debris mitigation, the potential for injury and loss of life is substantially reduced.

Blast-resistant design criteria define how much damage to a façade or structure is considered acceptable. At the lowest level of performance, the structure and façade may be designed to limit the number of casualties and allow evacuation of the building by preventing widespread, progressive structural collapse. The building itself may not be re-occupied without extensive repairs. This is typically referred to as a life-safety performance level.

At the highest level, it may be possible to design a structure or façade that would suffer no damage in the event of an explosion. This typically results in large, expensive structures and is generally restricted to specialist buildings, such as command and control centers with significant stand-off distances.

Effective risk-mitigation strategies optimize the delicate balance of protection benefits against the cost of implementation. A blast consultant can advise on the impact of designing for certain levels of protection and threat scenarios. There is often a desire for 100-percent protection, but this

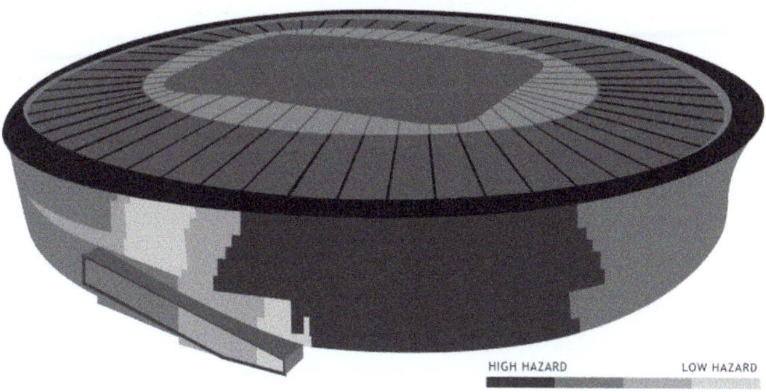

Figure 2.1 Damage Contours for a Façade Subject to Air Blast

is typically neither economically feasible, nor possible over the entire threat profile.

Development of site-specific performance criteria is complex and requires extensive experience. General performance levels are provided in several documents. Table 2.2 is extracted from ASCE 59–11.[1] Its principles are replicated in some form in most design guidance. The level of protection (LoP) corresponds to the expected performance of building elements under the blast loads generated from the DBTs and the corresponding impact upon building occupants. Where life safety is the desired performance criterion, a low LoP is generally adopted.

The descriptions shown in Table 2.2 provide a general summary of blast performance. For design purposes, quantitative member response levels must also be provided. Most sources use single-degree-of-freedom (see later sections) response limits, which state how far a structural member may displace before failing. Blast engineering is not codified, but there are some guidelines available. These are highlighted in Figure 2.2.

2.4 BLAST LOADING

The three key characteristics that define an explosion are:

1. It is near instantaneous—occurring in milliseconds.
2. It generates an intense release of energy.
3. It generates a compression wave in the surrounding medium.

The chemical and physical processes taking place during an explosion are highly complex. A number of excellent references[2–4] are available that explain these concepts from first principles.

When a combustible material undergoes a near-instantaneous decomposition, the process is known as detonation. Figure 2.3 shows the pressure–time relationship for such a blast. It generates a shattering shockwave in materials that are in contact with the explosive and creates a pressure wave if the expansion occurs in air or water. This pressure is of very high magnitude, over a very short duration (milliseconds).

Explosive gases do not tend to exhibit detonative behavior; instead, the combustion process is known as deflagration. This results in objects nearby experiencing a temporary increase in ambient **overpressures** much lower than a detonation, and over a slightly longer timescale, shown in Figure 2.4.

Table 2.2 Performance Definitions for Various Levels of Protection

Level of Protection	*Description of Potential Overall Structural Damage*	*Potential Door & Glazing Hazards*	*Potential Casualties*
Below Standards	*Severe Damage*	*High Hazard*	*Severe Casualties*
	Progressive collapse likely. Space in and around damage area is unusable	Doors and glazing will fail catastrophically and result in lethal hazards	Majority of personnel in collapse region suffer fatalities. Fatalities in areas outside collapsed area likely
Very Low	*Heavy Damage*	*Low Hazard*	*Heavy Casualties*
	Onset of structural collapse. Progressive collapse is unlikely. Space in and around damage area is unusable	Glazing will fracture, come out of the frame, and is likely to be propelled into the building, with the potential to cause serious injuries. Doors may be propelled into rooms, representing serious hazards	Majority of personnel in damaged area suffer serious injuries, with potential for fatalities. Personnel in areas outside damaged area will experience minor-to-moderate injuries
Low	*Low Unrepairable Damage*	*Very Low Hazard*	*Moderate Casualties*
	Progressive collapse will not occur. Space in and around damage area is unusable	Glazing will fracture and potentially come out of the frame, but at a reduced velocity; it does not present a significant injury hazard. Doors may fail, but	Majority of personnel in damaged area suffer minor-to-moderate injuries, with the potential for a few serious injuries, but fatalities are

Table 2.2 Continued

Level of Protection	*Description of Potential Overall Structural Damage*	*Potential Door & Glazing Hazards*	*Potential Casualties*
		they will rebound out of their frames, presenting minimal hazards	unlikely. Personnel in areas outside damage areas will potentially experience minor-to-moderate injuries
Medium	*Repairable Damage*	*Minimal Hazard*	*Minor Casualties*
	Space in and around damage area can be used and is fully functional after clean-up and repairs	Glazing will fracture, but remain in the frame and result in a minimal hazard consisting of glass dust and slivers. Doors will stay in frames, but will not be reusable	Personnel in damaged area potentially suffer minor-to-moderate injuries, with the potential for a few serious injuries, but fatalities are unlikely. Personnel in areas outside damage areas will potentially experience superficial injuries
High	*Superficial Damage*	*No Hazard*	*Superficial Casualties*
	No permanent deformations. The structure is immediately operable	Glazing will not break. Doors will be reusable	Only superficial injuries are likely

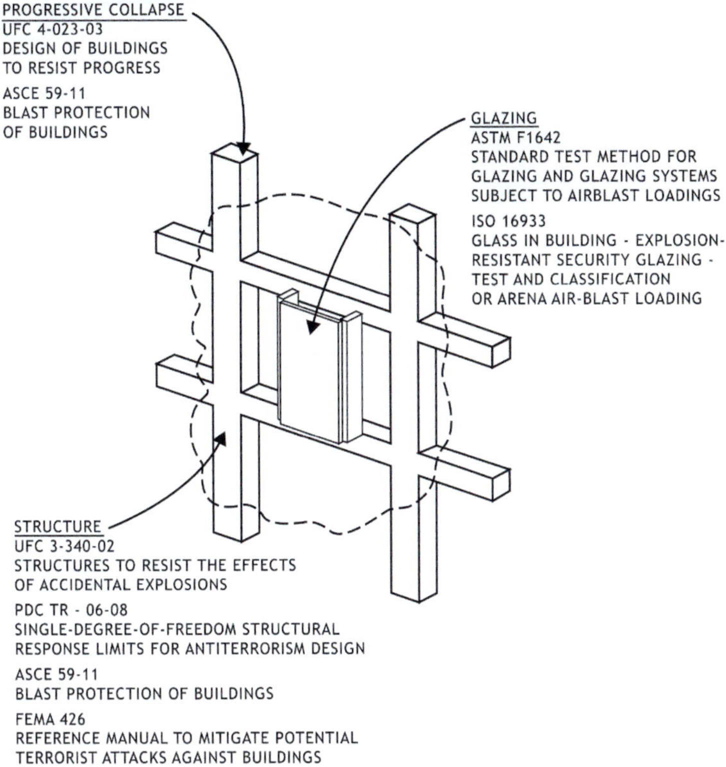

Figure 2.2 Diagram Illustrating Blast Resistant Guidelines for Façade and Structural Design

In addition to the expanding pressure wave, fragments are also thrown out. Blast-resistant structural design usually omits the effects of fragmentation. Fragmentation design shares many aspects with ballistic protection. However, combined blast and fragmentation protection can be tough to achieve economically.

2.4.1 Air-Blast Loads

Air-blast loads are often compared to seismic loads, but they are different in duration, magnitude, and the extents of loading, as illustrated in Figure 2.5. Seismic loads are cyclic and affect the entire structure, whereas blast loads are impulsive and affect a local region.

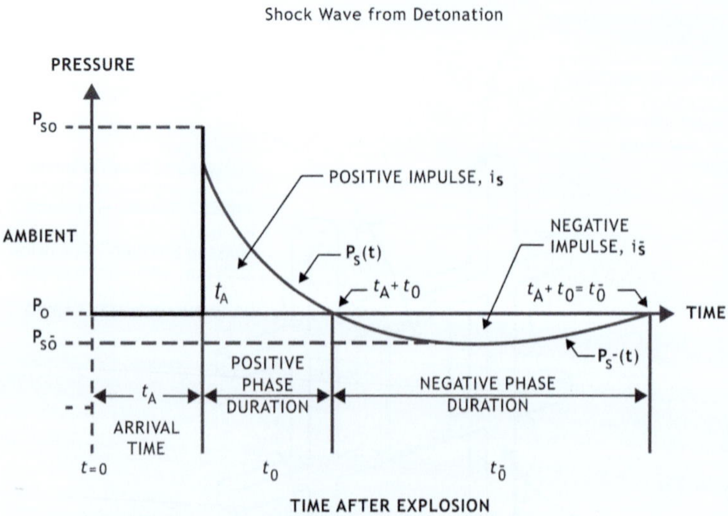

Figure 2.3 Shock Wave from Explosive Detonation

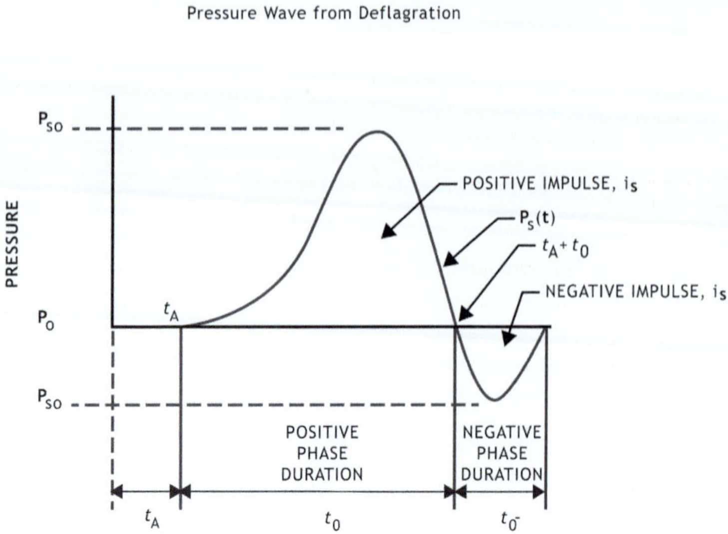

Figure 2.4 Shock Wave due to Deflagration

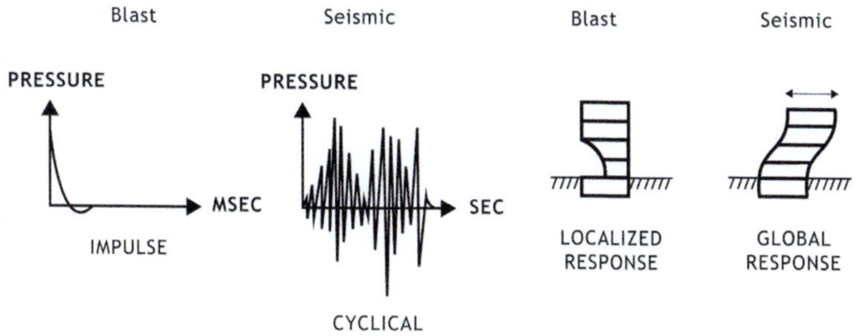

Figure 2.5 Blast versus Seismic Loading

As the expanding pressure wave encounters a rigid obstacle (such as a building), it applies a load to the obstacle (pressure). Note that blast pressures always act normal to the surface they are interacting with, irrespective of their direction of travel. As a blast wave passes over a building, it also generates an inward pressure load on the rear face, as shown in Figure 2.6.

The intensity of blast loads is dependent upon:

- the size of the explosive device (charge weight);
- the distance between the explosion and the building in question (range);
- the orientation and size of the building (angle of incidence and clearing);
- the surrounding environment (reflections);
- the type of explosive (TNT equivalence);
- the geometry of the explosive device (shaped charges).

Two design parameters are typically considered in design for blast loads: these are the **reflected pressure**, p_r, and the **reflected impulse**, I_r. The impulse is equivalent to the area under the pressure–time curve, as shown in Figure 2.3.

Typically, the negative phase is neglected, and the positive phase is idealized as a triangular distribution; therefore, the impulse energy = ½ P^*t_0, where t_0 is the duration of the blast load, as shown in Figure 2.7.

2.4.2 Internal Explosions

Internal explosions are a special case. As there are multiple surfaces, the pressure wave reflects multiple times and can combine into a larger pressure wave in a process called "shocking up." Additionally, the combustion gases from the explosion remain within the enclosed space, applying a near-constant pressure load, illustrated in Figure 2.8.

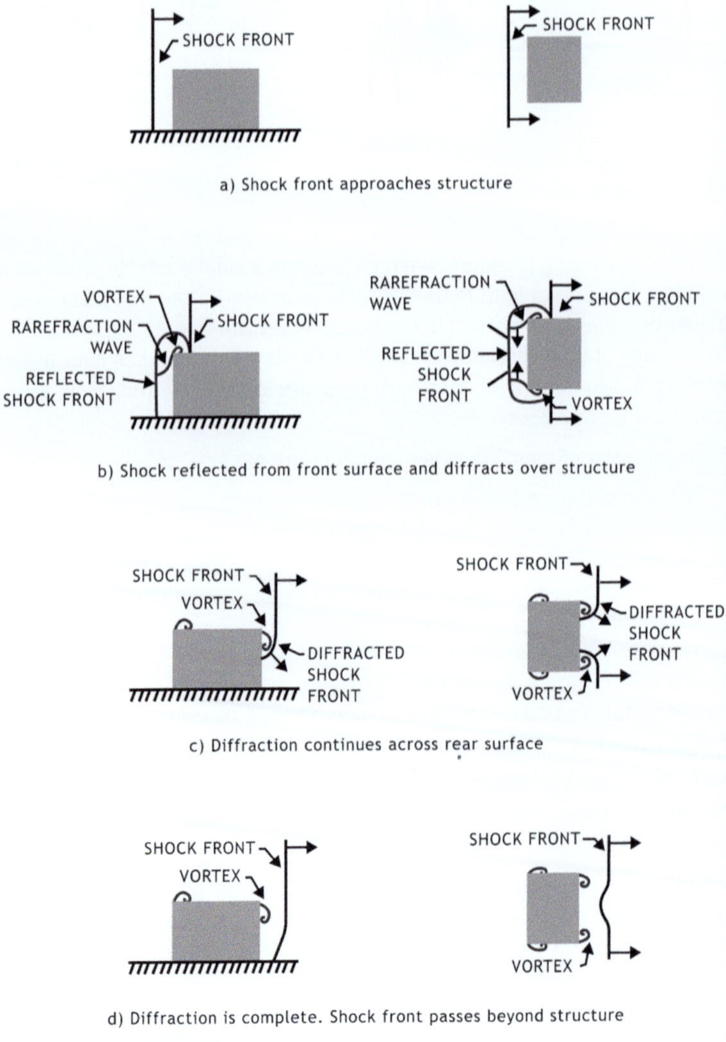

Figure 2.6 Shock Wave Traveling around a Building

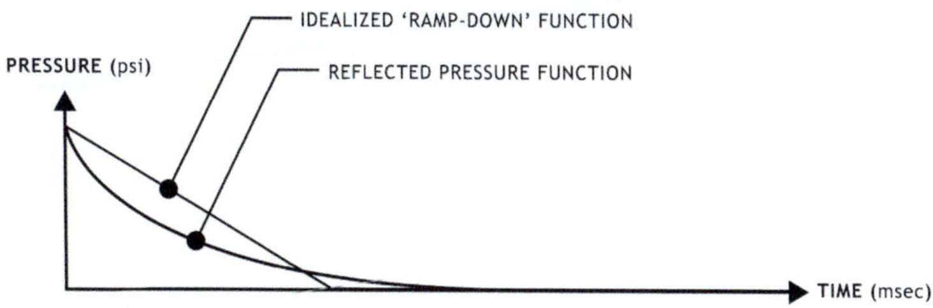

Figure 2.7 Idealized Shock Wave

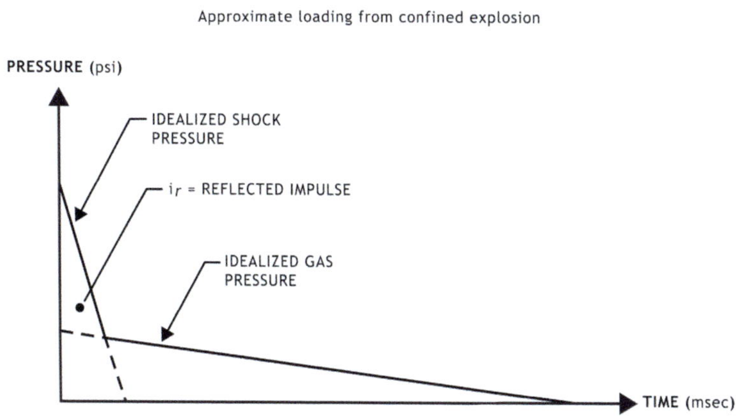

Figure 2.8 Approximate Loading from Gas Explosion

So that this load can be relieved, some form of venting to the external atmosphere is required. The size and location of the vent relative to the explosion are important for its effectiveness: the larger and closer it is, the better. Note that a blast vent does not prevent the initial blast load, but it does allow the reflected waves and combustion gases to escape, relieving the loads on internal surfaces.

2.5 SITE AND BUILDING LAYOUT

2.5.1 Stand-off

Stand-off is the distance between an explosive threat and the building façade and can only be guaranteed by the use of **vehicle**-impact-rated **security barriers**,[5] such as bollards, walls, and planters, as shown in

Figure 2.9. These can be very expensive. For sites with sufficient space, utilizing landscaping to provide berms, ditches, and ponds can be effective, more attractive, and less costly than more conventional bollard systems.

Blast loads are typically governed by the following **scaled distance** relationship: $Z = R/W^{1/3}$, where Z is the scaled distance, R is the stand-off distance in feet, and W is the charge weight in pounds of TNT.

In this cubic relationship, a doubling of the distance between the bomb and a building is equivalent to reducing the bomb weight by a factor of 8. Early consideration of site layout and building location within the site can pay huge dividends in terms of maximizing distance from vehicle threats and reducing the cost of blast enhancements.

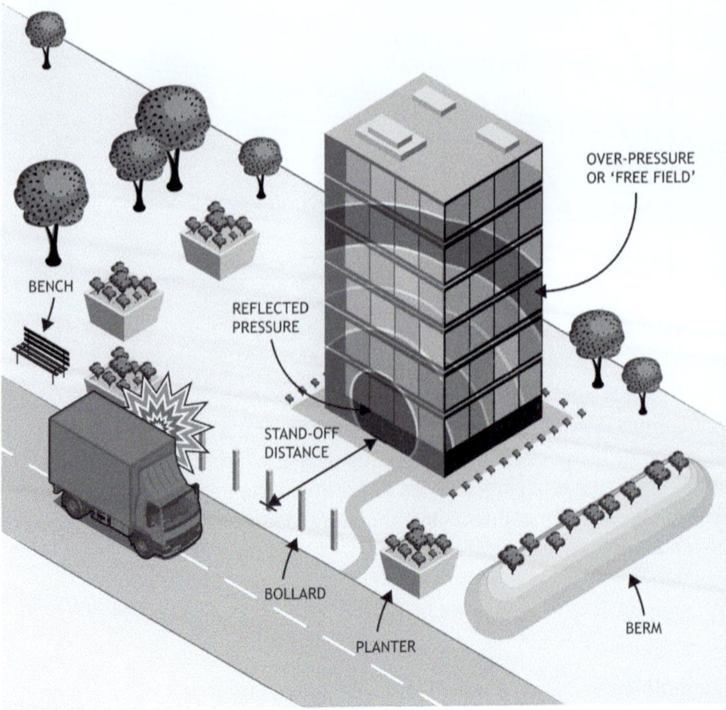

Figure 2.9 Illustration of Stand-off Distance

2.5.2 Parking

Providing parking within the building footprint, particularly under occupied areas, represents a significant increase in risk. For urban sites, there may be no other options, and so the risk needs to be managed through a balance of operational, electronic, and physical security measures. The consequence of a large vehicle explosion within a building is potentially catastrophic, and it is not typically economic to provide blast mitigation alone.

2.5.3 Layout

Protection that is more cost-efficient can be provided to building occupants by advantageous use of space planning within the building. This separates the high-risk areas (lobby, loading docks, mailrooms, or parking garage) from critical utilities and occupied space. Figure 2.10 summarizes some architectural planning recommendations.

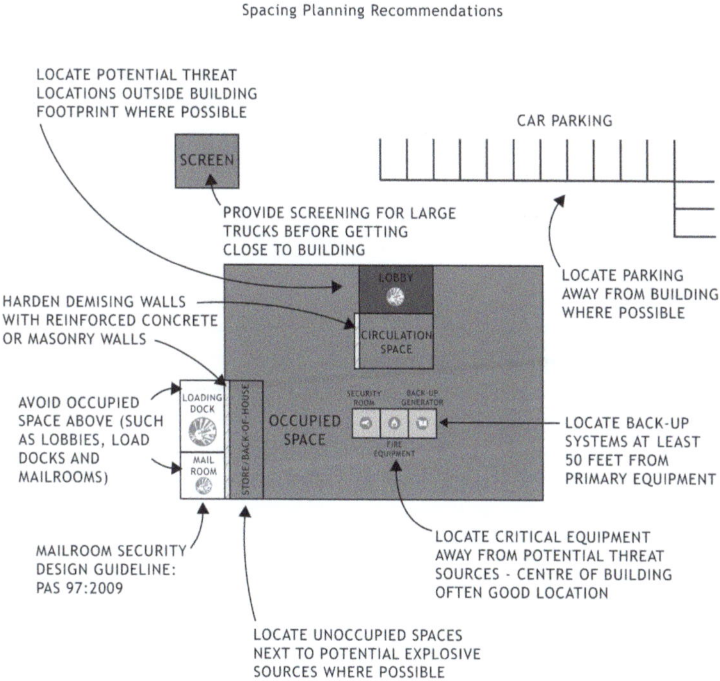

Figure 2.10 Architectural Planning Recommendations

2.6 FACADE CONSIDERATIONS

2.6.1 Hazards

The primary design objective for the exterior envelope of most standard buildings is to exclude blast pressures and mitigate the hazard of flying debris, to reduce casualties and business disruption and facilitate rescue and evacuation efforts. Whereas the majority of fatalities in terrorist bombings are attributable to structural failure, most non-lethal injuries are the result of façade failures. Of these, the failure of glazing is responsible for the overwhelming majority.

Glass is a brittle material and forms highly hazardous fragments when subjected to air-blast loads. It fractures many hundreds of yards (meters) from the explosion source, magnifying the number of casualties of an explosion. Glazed façades are, therefore, the primary focus of this section. However, many of the principles outlined here equally apply to non-glazed façades.

2.6.2 General

The performance of a building façade under blast loads is governed by the weakest component in the load path back to the primary structural system, beyond which the mass and strength of the primary structure are generally significant enough to resist the applied loads.

The most important aspect of designing façades for blast loads is to provide robust details and clear load paths back to the primary structural frame. It is best for façade systems to span floor to floor, rather than horizontally, to avoid loading on structural columns in bending. Such designs and details may conflict with architectural and other façade performance objectives, and so early engagement between the blast engineer and other design disciplines is vital.

2.6.3 Good Practice Measures

Where an explicit engineered level of blast resistance is not required, but some additional resilience is desired, consider the suggestions provided in Figure 2.11 and adopt a balanced design.

2.6.4 Balanced Design

A **balanced design** approach maximizes the protection provided by the façade. This means that a hierarchy of failure is established to ensure that the framing allows the maximum capacity of the cladding to be developed, and, subsequently, the connections allow the maximum flexural capacity of

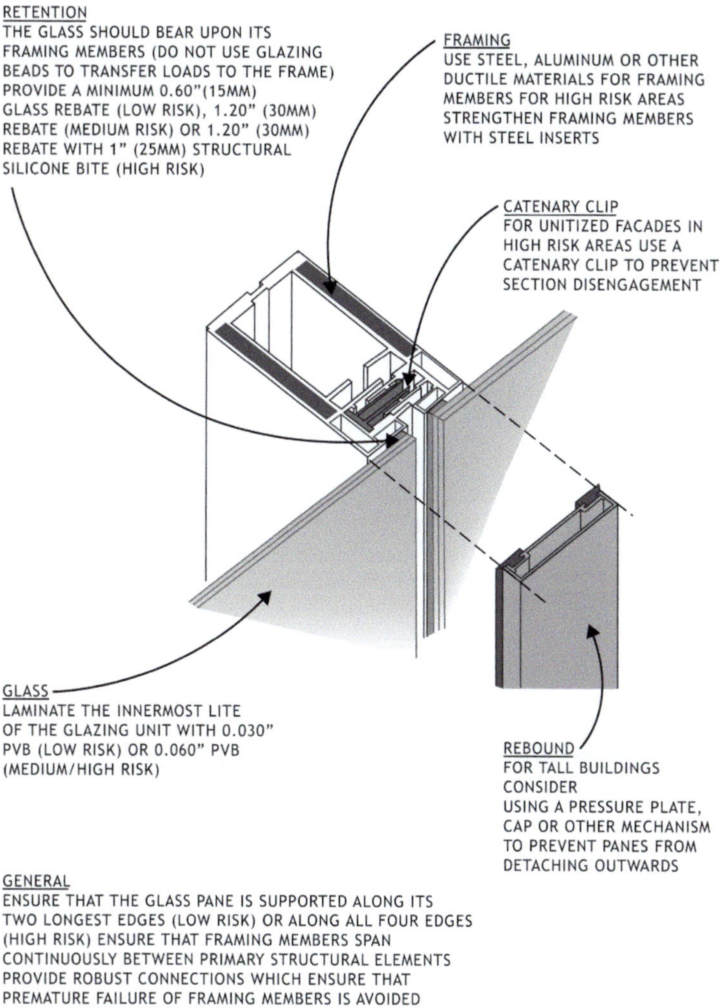

Figure 2.11 Good Practice Recommendations for Reducing Debris due to Air Blast

the framing to be developed. This ensures that the failure sequence is controlled, and a sudden failure of the whole system is avoided.

2.7 STRUCTURAL CONSIDERATIONS

2.7.1 General

Conventional structural design loads are defined according to the building's intended use and location. The structure is then designed to resist these loads. To limit the likelihood of structural failure, the design loads and structural resistance are modified by factors. This concept underlies most modern load and resistance factor design (LRFD) or limit state design codes.

In blast-resistant design, the size of the explosive charge and its location are variable. As a result, there are often multiple design scenarios, rather than one single design load. The consequence of each of these possibilities needs to be determined in order for the overall performance of the building to be assessed. Where necessary, enhancements are made to ensure that the relevant performance criteria are achieved.

Unlike earthquakes, blast loading is usually a local rather than global design issue. It is the damage to individual members or assemblies that determines whether a building survives a blast event.

A structure with good performance under extreme loads should have the following:

- Strength—Critical members have the ability to resist larger than normal loads, increasing their resistance.
- Ductility—Maximize the strain energy that a structure may absorb.
- Redundancy—Have alternate load paths so that loads can be redistributed in the case of structural failure.

2.7.2 Good Practice Measures

Figure 2.12 illustrates good practice structural design measures to be adopted within a building for increased resilience. Providing measures throughout is ideal, but, where not possible, begin by focusing on critical structural elements and those located in publically accessible areas. Start from the highest risk elements and work through to the lowest.

2.7.3 Disproportionate and Progressive Collapse

By far the most important aspect when a building is designed is protection against progressive or disproportionate collapse.

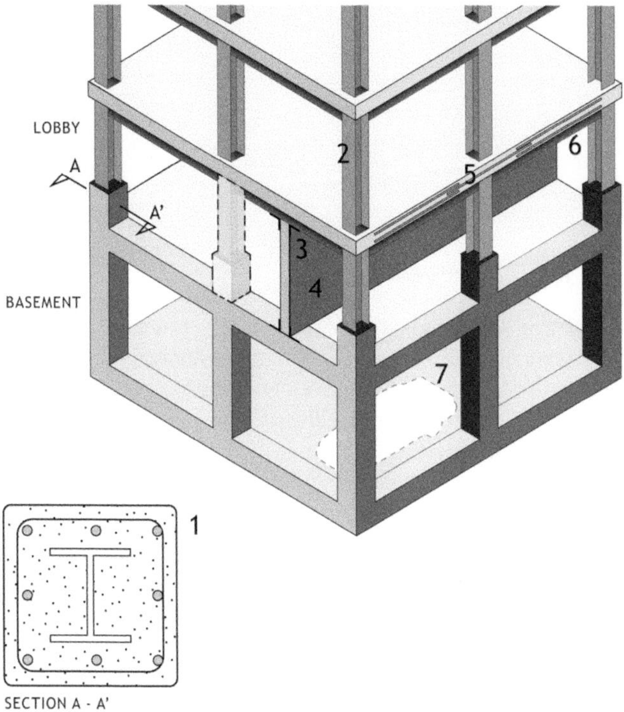

1 STEEL COLUMN HARDENED FOR PBIED WITH CONCRETE ENCASEMENT ~5FT

2 USE DUCTILE STRUCTURAL MATERIALS SUCH AS REINFORCED CONCRETE OR STEEL. UTILIZE SEISMIC DETAILING

3 PROVIDE ROBUST CONNECTION INTO SLAB

4 PROVIDE ROBUST INTERNAL PARTITIONS (REINFORCED GROUTED CMU OR REINFORCED CONCRETE) TO SEPARATE UNSECURE SPACE FROM OCCUPIED OR CRITICAL AREAS

5 PROVIDE SYMMETRIC REBAR IN SLABS TO RESIST BOTH UPWARD AND DOWNWARD FORCES

6 AVOID TRANSFER GIRDERS ABOVE UNSECURE SPACE

7 TRY TO AVOID PARKING UNDER THE BUILDING. IF NECESSARY, SCREEN VEHICLES AND LIMIT PARKING AND PARKING ACCESS TO MOST VULNERABLE AREAS (NEAR CORE AND PRIMARY COLUMNS)

Figure 2.12 Structural Good Practice Recommendations for Increased Resilience

Disproportionate collapse is defined by the ASCE[6] as "a collapse that is characterized by a pronounced disproportion between a relatively minor event and the ensuing collapse of a major part or whole of a structure." **Progressive collapse** is defined in the same document as "a collapse that commences with the failure of one or a few structural components and then progresses over successively affected other components."

Typical approaches for resistance to disproportionate or progressive collapse prevention are (1.) adding ties, (2.) analyzing the structural response to removed elements, and (3.) hardening critical elements.

Horizontal and vertical ties enable catenary action to develop, hold columns up, and allow loads to be redistributed around lost members. Tie force methods are probably the commonest method for preventing building collapse, but they are also widely misused or applied outside their range of applicability. The intention is that loads that were previously carried by columns can be turned into axial loads as an element is suddenly removed, as illustrated in Figure 2.13. This can only occur when the connections in the structure are able to deform (rotate) and maintain their integrity. For modern connection types (not designed for earthquakes), rotations exceeding 2–3° will usually cause a connection failure. Code-based tie force methods should only be used for buildings with modest spans and regular, rectangular grids. Outside this, they may still be used, but greater analysis is required (alternate path method).

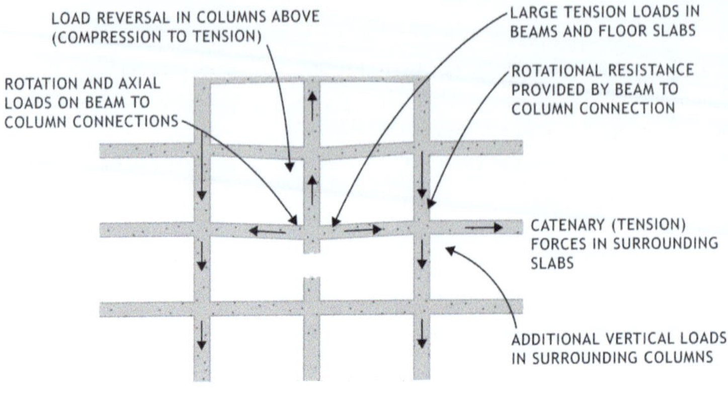

Figure 2.13 Illustration of Tie Force Methodology

Element removal analysis (also known as alternate path analysis) involves removing elements (one or more, depending upon the threat) in sequence and determining the necessary structural redundancy to redistribute loads. Appropriate methods for alternate path analysis can be found in UFC 4–023–03[7] and ASCE 59–11[8] and include static, dynamic, linear, and nonlinear methods.

Critical elements should be hardened so that their loss due to an extreme event is unlikely. However, this makes the design threat-dependent (i.e., good for a given load size), whereas the column removal is threat-independent up to a point.

A blend of all these methods will usually yield the most cost-effective and efficient collapse-prevention strategy.

2.7.4 Structural Blast Design

During design for blast loading, structural elements are checked for their relevant modes of failure. For example, concrete modes of failure are provided in Figure 2.14.

Flexure is a ductile and, hence, desirable failure mode. For a member to yield in flexure, it should:

- be plastically compact (steel);
- have connections capable of developing the yield capacity;
- be resistant to shear and breach failures.

Seismic detailing is very effective in achieving this behavior.

2.7.5 Services

Some projects may require the building services (water, power, HVAC) to withstand the effects of a blast. Relatively little guidance is available on acceptable levels of damage to services. Current practice is to ensure that risers and ducts remain elastic under blast loads, and that the services themselves are isolated from excessive accelerations (e.g., not fixed to blast-resisting walls). Exterior services such as backup generators can be protected by being encased in a concrete enclosure.

2.8 ANALYTICAL APPROACHES

It is common for structural engineers with few previous encounters with blast-resistant design to typically adopt one of two paths when confronted with a problem involving blast loads:

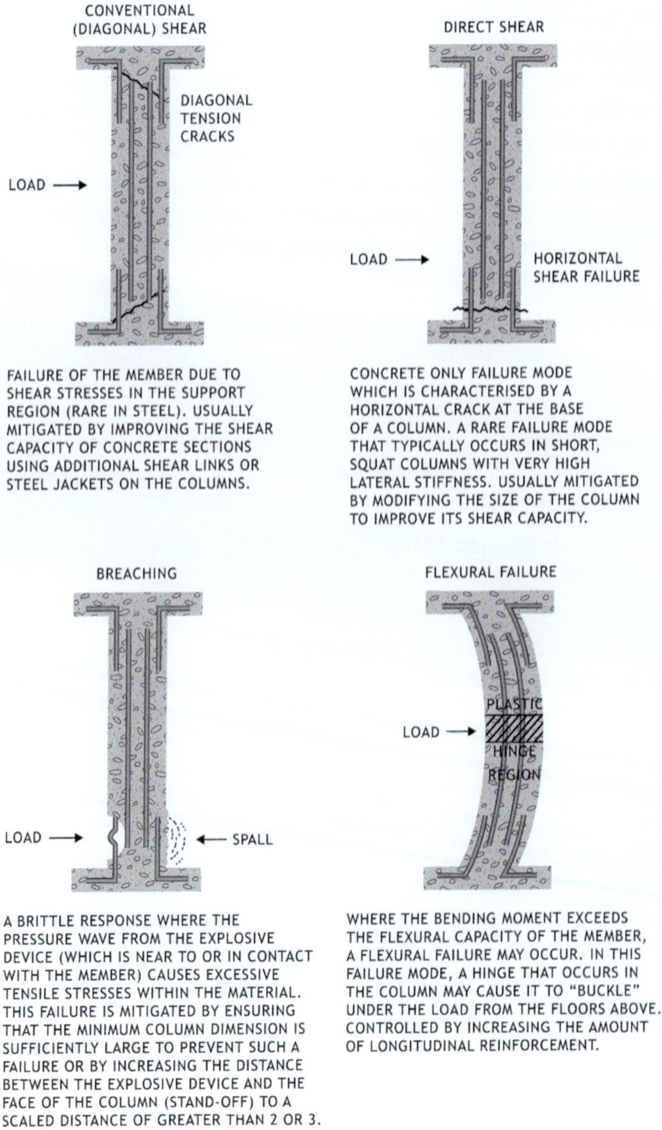

Figure 2.14 Structural Modes of Failure (Concrete)

1. Apply the blast pressure (potentially tens of pounds per square inch/hundreds of kilonewtons per square meter) as a static load to the structure.
2. Ask for an equivalent static load.

The first results in an excessively robust but expensive structure. The second is possible but requires knowledge of the dynamic characteristics of the element in question. It will also result in an excessive design, as design in the **elastic** range does not account for the nonlinear response (energy dissipation capacity) of the materials.

The simplest technique for determining the response of individual elements to dynamic blast loads is the single-degree-of-freedom (SDoF) method. The element is simplified as a mass-spring system, where member characteristics are represented by the spring. Despite its limitations, SDoF is widely used, as it offers designers a quick, simple method to conduct blast analyses. Output usually corresponds to the damage levels defined in the project performance criteria. Owing to its short analysis time, SDoF can be used to generate pressure–impulse (P–I) curves, shown in Figure 2.15. These graphically represent the combinations of pressure and impulse that bring about the same level of damage.

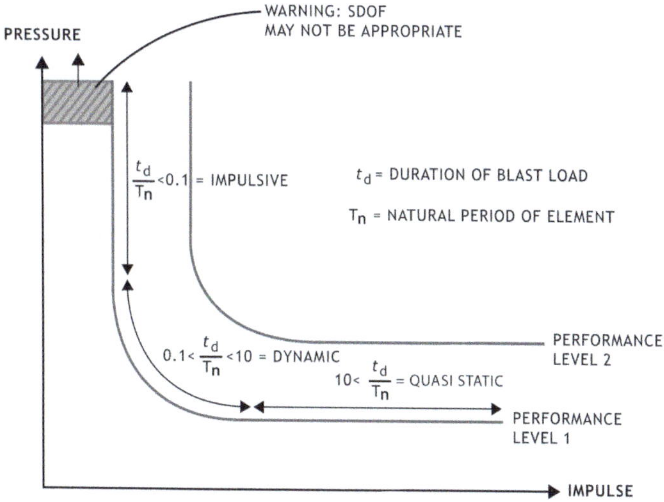

Figure 2.15 Pressure–Impulse Curve for Two Performance Levels

The P–I curve can be broken down into three distinct regions: impulsive, dynamic, and quasi-static.

In the impulsive region, damage is controlled exclusively by the blast impulse. An impulsive response typically occurs for small, close-in explosive devices and instances where the member is highly flexible or has very large mass. In this region, the best way to improve performance is to add mass. A system is impulse-controlled when $t_d/T < 0.1$, where t_d is the duration of the blast load, and T is the natural period of the member in question.

In the quasi-static range, the dynamic properties of the member (mass and stiffness) are largely irrelevant. Member strength is what counts. Quasi-static response is generally limited to scenarios where the blast is large and a long way from the building (e.g., petrochemical or nuclear explosions). In this region, the best way to improve performance is to strengthen the member. A system is quasi-static or pressure controlled when $t_d/T > 10$.

In the dynamic region (systems where t_d/T is between 0.1 and 10), the damage to a structural element is a function of both pressure and impulse. Here, the mass, stiffness, and strength of members all contribute to its blast capacity.

Most modern engineering software is able to handle geometric, material, and loading nonlinearity at the global and structural element level and is well suited for blast analysis of materials such as steel and reinforced concrete. To analyze more complex materials, such as fiber-reinforced composites, or to look at material response at the smaller scale, you need to use **finite element packages**, such as LS-DYNA™ or Autodyn™.

Given the large number of unknowns in blast engineering, it is better to choose the simplest analytical method that gives you a good degree of certainty. It is prudent to establish upper and lower bounds for damage using a simple tool, rather than rely on more complex models requiring ever greater input data.

2.9 DESIGN INTEGRATION PROCESS

Blast engineers are involved throughout all phases of the project, starting with due diligence through construction administration. A brief summary of design stages and design input/deliverables is provided in Table 2.3. So that costs can be minimized, it is essential the blast engineer be integrated into the design process at an early stage. Thorough coordination

Table 2.3 Blast Collaboration for Each Design Stage

Design Phase	*Input Required*	*Blast Consultant Scope*	*Deliverables*	*Interfaces*
Pre-concept	Threat & risk assessment DBT determination Owner performance expectations	Criteria development Site selection/ building selection (for lease or buy) Threat & risk assessment	Blast basis of design information Site/building selection recommendations report	Tenant Client/owner Security consultant Architect
Concept	Architectural & structural concept information, site plan, security strategy	Advice on spatial planning, building massing, façade and material types, structural system selection	Report/ presentation providing good practice guidance Agreed stand-off distances	Architect Structural Engineer MEP engineer, Security consultant Client/owner Cost consultant Façade engineer
Schematic	Design drawings, structural dead and live loads, structural material properties and preliminary sizes, façade system types	Preliminary analysis of structural/ architectural options to inform costing/design decisions	Report outlining preliminary analysis results and recommendations	Architect Input to cost plan Structural engineer MEP engineer

Table 2.3 Continued

Design Phase	*Input Required*	*Blast Consultant Scope*	*Deliverables*	*Interfaces*
				Security consultant Client/owner Cost consultant Façade engineer
Detailed	Design drawings, structural dead and live loads, structural material properties and sizes, façade system types with supporting systems identified and designed for wind/gravity loads	Analysis of typical structural columns and beams (and possibly slabs) Proof-of-concept façade analysis Development of blast-resistant façade specifications	Identify members requiring enhancement to meet performance requirements, provide connection/reaction forces to SEOR Report outlining findings Draft performance specifications for structural/façade systems	Architect Structural engineer MEP engineer Security Consultant Client/owner Cost consultant Façade engineer/contractor
Construction documents	Design drawings, structural dead and live loads, structural material properties and sizes,	Analysis of atypical structural/façade systems, progressive	Identify members requiring enhancement to meet performance requirements, provide	Architect Structural engineer Security consultant

Table 2.3 Continued

Design Phase	*Input Required*	*Blast Consultant Scope*	*Deliverables*	*Interfaces*
	façade system types with supporting systems identified and designed for wind/gravity loads	collapse analysis (if required) Review drawings to check blast recommendations incorporated	connection/reaction forces to SEOR Report outlining findings Final performance specifications for structural/façade systems	Client/owner Cost consultant Façade engineer/contractor
Construction administration	Contractor drawings and calculations in conformance with performance specifications	Review contractor submittals related to blast requirements—coordinated through lead designer	Submittal review documents Site supervision for complex, high-performance systems	Architect Structural engineer Contractor Façade engineer/contractor Cost consultant

throughout the process is essential, so that challenges can be resolved quickly, and the engineer is aware of any changes to the design that may affect their analysis assumptions.

2.10 SUMMARY

Blast design is crucial for the protection of human life and critical buildings. Blast performance can range from very low levels of protection after a small blast (most buildings today) to fully operational after a major event. Threats may range from backpack to semi-sized charges. The complexity and risk require sound engineering from specialty blast engineers who can guide the design team to an economical and safe solution.

NOTES

1. AISC. *Blast Protection of Buildings*, ASCE 59–11 (Reston, VA: American Society of Civil Engineers, 2011).
2. P.D. Smith and J.G. Hetherington. *Blast and Ballistic Loading of Structures* (Boca Raton, FL: CRC Press, 1994).
3. CCPS. *Guidelines for Vapor Cloud Explosion, Pressure Vessel Burst, BLEVE and Flash Fire Hazards* (New York: CCPS, 2010).
4. C.E. Needham, *Blast Waves* (New York: Springer, 2010).
5. ISO, *Vehicle Security Barriers—Part 1: Performance Requirement, Vehicle Impact Test Method and Performance Rating*, IWA 14–1:2013 (Vernier, Switzerland: ISO, 2013).
6. U. Starossek and M. Haberland. "Disproportionate Collapse: Terminology and Procedures," *Journal of Performance of Constructed Facilities* 24, no. 6 (December 2010): 519–528.
7. DoD, *Unified Facilities Criteria 4–023–03 Design of Buildings to Resist Progressive Collapse* (Washington, D.C.: Department of Defense).
8. ASCE, *Blast Protection of Buildings*, ASCE 59–11 (Reston, VA: American Society of Civil Engineers, 2011).

Sustainable Structures

Chapter 3

Saglinda H. Roberts
and Rob Fleming

At the core of all architecture lies structure. At the core of civilization's well-being lies **sustainability**. Structural design and sustainable design are inseparable because they both serve to support our pursuit of a sustainable building in the short term and a sustainable future in the long term. Your role in helping to shape a sustainable future is much broader and deeper than you might first assume.

The responsibilities of a structural engineer stretch beyond the short-term needs of protecting human safety in built structures. They extend to protecting the long-term needs of society in general. Buildings are the number one driver of **carbon emissions**,[1] not just from the operation of lighting, heating, and cooling, but more directly from the embodied energy use of the very structures you design. Concrete, steel, and wood all possess a **carbon footprint** that, when used irresponsibly, release millions of metric tons of **CO_2** gases into the atmosphere. The global impacts of this are well known. According to the Intergovernmental Panel on Climate Change,[2] increasing temperatures, sea-level rise, loss of **biodiversity**, ocean **acidification**, and extreme weather are the result of increased levels of CO_2 and other greenhouse gases in the atmosphere. There is no doubt that future generations will bear the burden of dealing with the escalation of these problems.

Sustainability, then, is really the expression of empathy for those people not yet born. The concept was first expressed by the Brundtland Commission from the United Nations in the seminal document *Our Common Future*, where it is stated that, "Sustainable development is the development that meets the needs of the present without compromising the needs of future generations to meet their own needs."[3] The use of the word needs is critical, as it signals a shift in design of the built environment away from the styles, status, and excess (wants) to a new model of design that focuses on meeting the long-term needs of society.

Figures 3.1–3.3 reflect the need to change and the critical role **sustainable design** plays in being the driver of change. In Figure 3.1, the vectors of available **resources** and demand for them are on a collision course, leading us into an unsustainable future.[4]

Alternatively, Figure 3.2 depicts a possible future shift where sustainable design becomes regenerative, ensuring the ability of future generations, not only to meet their needs, but to thrive.

We are currently in what some term a **green plateau**, where buildings are now mitigating their damage to the environment but not helping to regenerate **ecosystem** services, as illustrated in Figure 3.3.

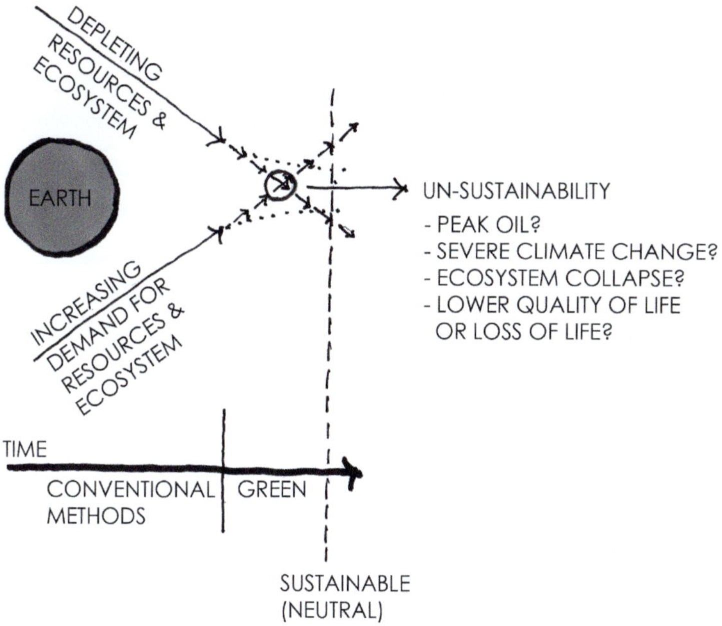

Figure 3.1 Collision Path of Resources and Demand

Source: Adapted from the Natural Step Program

3.1 FRAMEWORKS FOR AUTHENTIC SUSTAINABILITY

One of the great barriers to the widespread adoption of sustainable design strategies is the lack of a commonly accepted framework. Authentic sustainability is far more than LEED or any rating system, or materials or energy sources alone. A new regenerative model of design can only be accomplished with **holistic** frameworks, based on empathy and collaboration across the disciplines. It acknowledges the interrelated, reciprocal relationship of the site, structure, client, community, and **urban context**, along with the professions, and seeks a multi-lens approach.

3.1.1 Integral Theory

Integral theory was first applied to the architectural field by Mark DeKay in his book *Integral Sustainable Design*.[5] It expresses the belief that all human understanding and perception can be broadly categorized into four

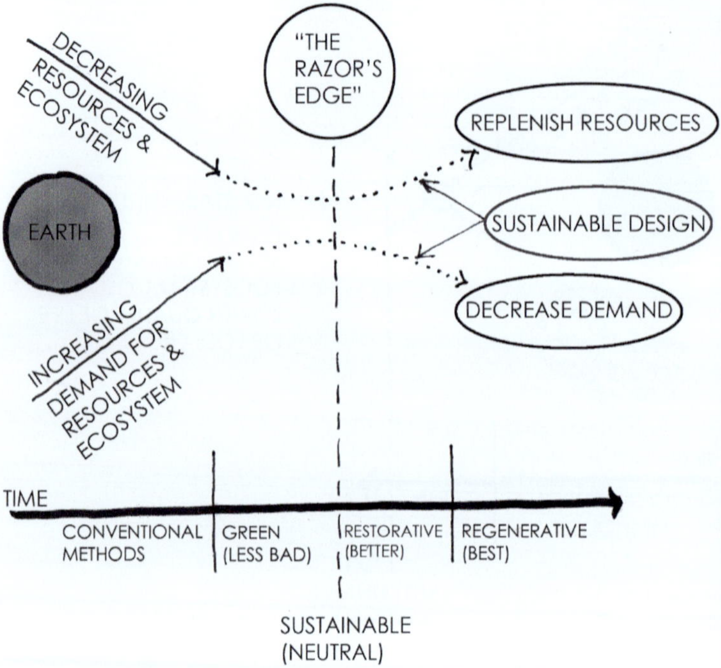

Figure 3.2 Possible Diverging Path of Resources and Demand

Source: Adapted from the Natural Step Program

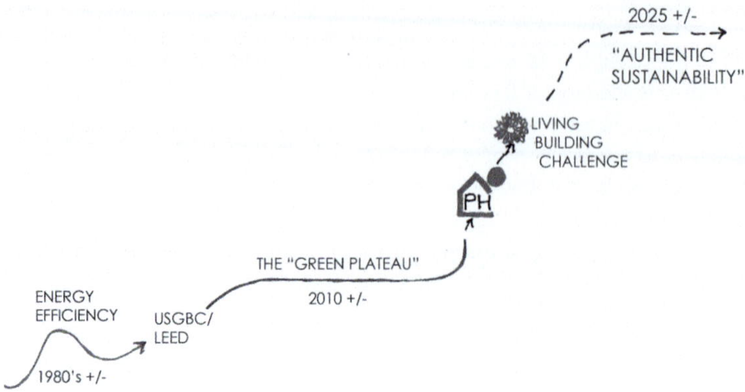

Figure 3.3 The Green Plateau of Sustainable Design

main areas of experience or beauty, behavior or function, cultural connections, and systems or **ecological** perspective.

The holistic, integral analysis process helps identify deeper issues, allowing a synthesis of the dynamic goals of aesthetics, function, sustainability, systematic functioning, and cultural connections. Use of this methodology at the beginning of a project allows all stakeholders to have a comprehensive understanding of the client, occupant, societal, ecological, and cultural needs of a project.

This multi-lens approach allows us to see sustainable design, not just as a tool to achieve higher energy performance, but also as a vehicle to achieve higher levels of social equity, beauty, and systems integration. DeKay's four-quadrant grid, shown in Figure 3.4, shows the design directives in each perspective.

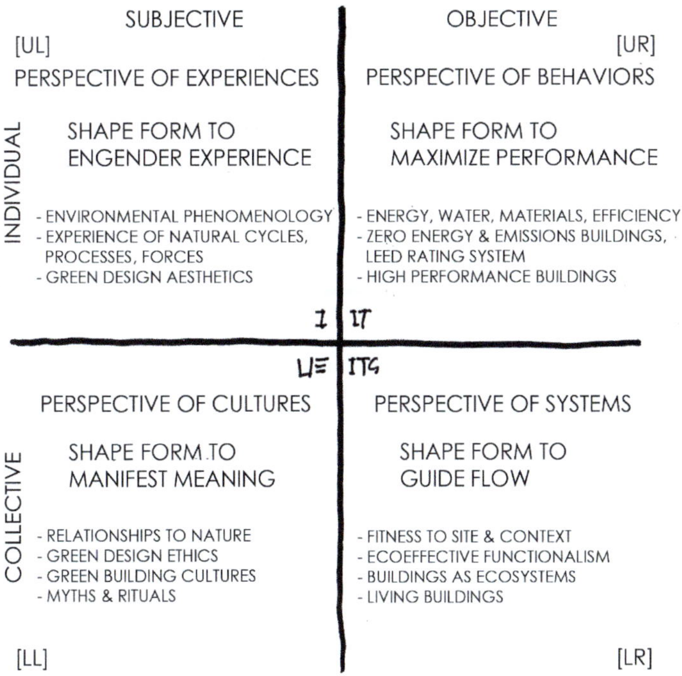

Figure 3.4 Integral Theory Grid

Source: Mark Dekay. *Integral Sustainable Design: Transformative Perspectives*[6]

The structural engineer, like other design team members, can use this framework when developing a comprehensive approach to sustainable design on every level. Structure elements can become a vehicle for achieving higher levels of beauty, deeper ethical foundations, systems integration, and increased energy performance with cost-efficiency.

3.1.2 Worldview

Dr. Ashraf Salama, founding head of the Department of Architecture and Urban Planning at Qatar University, writes, "the built environment [is] a two-way mirror . . . it conveys and transmits non-verbal messages that reflect inner life, activities, and social conceptions of those who live and use the environment."[7] As we briefly examine the evolution of structures, we see, not just a series of technical advancements, but also—perhaps more importantly—an expression of a constantly evolving reflection of society's view of its relationship to the Earth and its resources, expressed in architectural form and site relationship. These worldviews go on to affect all the decisions of a society, but specifically sustainable design initiatives and strategies:

- **Hunter-gatherer**—The so-called "primitive hut" was actually a carefully considered tension structure, made from local materials that maximized volume and minimized necessary materials. Hunter-gatherers believed themselves to be an integral part of the Earth, and their use of resources reflected that.
- **Agriculture**—In the age of agriculture, people made homes out of whatever was available. Pozzolanic cement was developed, not out of an intellectual scientific process, but from the trial-and-error process of manipulating locally available volcanic ash into a portable and moldable structural material. Monuments took on great stature because of the load-bearing capabilities of locally available stone. People were able to go beyond subsistence farming and store food; nature started to be seen as a resource or commodity.
- **Early Age of Industry**—The great domes and flying buttresses emerged as the ability to design and build larger and larger spaces developed, fulfilling an important social and cultural need. Science sought answers for the wonders of the world, causing humans to see themselves as separate from and above the Earth.
- **Late Age of Industry**—The use of steel became widespread, allowing more-functional space and the ability to build higher structures using a smaller footprint. The Industrial Revolution was in full swing, and

society was focused on production. Society saw nature as an endless resource at its disposal.

- **Age of Information**—Greater access to information caused the world to see and feel the impacts of our activities on the planet. Pollution, fossil fuel depletion, loss of biodiversity, and **deforestation** were impacting the ecological systems upon which we depend. Society became more aware of the ecological impacts of our materials selections.
- **Integral**—Mass timber, straw bale, and rammed earth, along with organic forms of architecture, confirm that society is moving towards a more integral relationship with nature. Humanity is beginning to reconnect to the natural world and find inspiration in its forms and processes.

An integral relationship to the world is not new, but it is critical for a truly sustainable approach. When we acknowledge and believe in our integral connection to the world, we view all our decisions in a different light and move closer to truly regenerative sustainability.

3.2 CHANGING ROLE OF THE STRUCTURAL ENGINEER

3.2.1 Integrative Design Process

It may not seem like the structural engineer has much of a chance to influence the sustainability or the carbon footprint of a building. In the traditional design process, the structural engineer is brought in very late to the game. He or she has little choice in making suggestions to reduce the carbon footprint of the building or change its form. The current linear model of design first, add structure later is giving way to a new model for developing design called the **integrated design process** (IDP), illustrated in Figure 3.5.

IDP is a **collaborative design** process involving all stakeholders at the very beginning of the design process. Design directives, goals, and areas of concern are expressed by all parties at the beginning of the process. This creates an environment of collaboration, enabling strategies to be explored and vetted through a multi-lens perspective and steered towards long-term, sustainable design. Having structural engineers involved is crucial.

IDP acknowledges that buildings and societies are better when we work collaboratively and evaluate design through a multi-lens perspective, to create and vet the best solutions. When used early in the process, IDP allows for dynamic **value engineering** and sustainable design strategies,

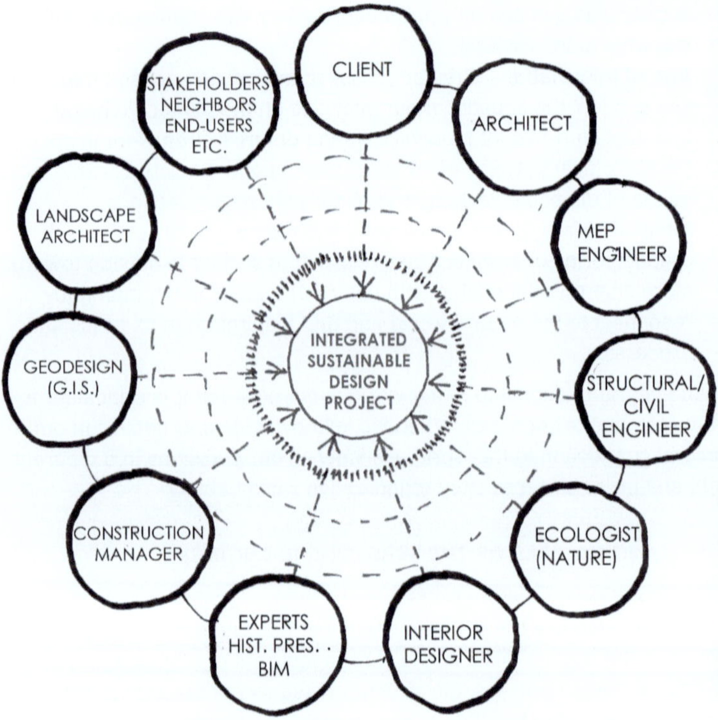

Figure 3.5 Interdisciplinary Model of the Integrated Design Process

Source: Design Education for a Sustainable Future, courtesy Routledge[8]

long before final decisions are made, thereby leading to greater efficiencies and less frustration in the process.

Collaborative involvement of engineers in the design process is not new. Dankmar Adler, an engineer, partnered with architect Louis Sullivan to form one of the most dynamic interdisciplinary design teams in modern history. The Auditorium by Adler and Sullivan reflected the best that each had to offer. The line between disciplines was blurred, and the result was better than either of them could have created separately.

3.2.2 The Future Landscape of Design

The professional landscape for a structural engineer is changing rapidly. As emerging professionals, they have the opportunity to bring this type of working model to their projects. Sustainable design is much broader and far deeper than one would first suspect. In the rest of the chapter, we will touch on several important basic areas. This chapter is meant to be a launching point for a lifetime of learning, rather than a definitive and comprehensive treatise on structural engineering and sustainability.

3.3 LIFE CYCLE ANALYSIS

To fully understand which type of structural material is the most sustainable in each specific situation, it is vital to look at the whole process of creating each material. This process is called **life cycle analysis** (LCA) and measures **embodied energy**. The process examines each step of manufacturing, from resource extraction through to final product delivery, including transportation, construction, and building maintenance. It also considers life expectancy, end-of-life disassembly, and reuse or disposal (Figure 3.6).

LCA is used to analyze the wide-ranging **environmental footprint** of a building—including aspects such as energy use, **global warming potential**, habitat destruction, resource depletion, and toxic emissions. We also need to examine how the community and surrounding ecosystem will be affected aesthetically, physically, and socially, even though there may not be facts or numerical figures to put on a spreadsheet.

LCA is one tool to aid understanding of the complexity of sustainability and shape our design decisions. For structural systems, the LCA is best done during the pre-design phase.[9]

Other benefits of utilizing LCA, as cited by the American Institute of Architect's study, are:[10]

1. choosing between building design or configuration options;
2. choosing between building structural systems, assemblies, and products;
3. reducing environmental impacts throughout a building's life cycle;
4. improving the energy performance of the entire building;
5. mitigating impacts targeted at specific environmental issues.

This integrative analysis and design process should take place very early, with all professions involved. As an emerging structural designer, you can introduce this new process.

3.4 AESTHETICS AND SUSTAINABILITY

Lance Hosey, in his book *The Shape of Green*, states: "if it's not beautiful it is not sustainable."[11] Stewart Brand, in *How Buildings Learn*, observed that

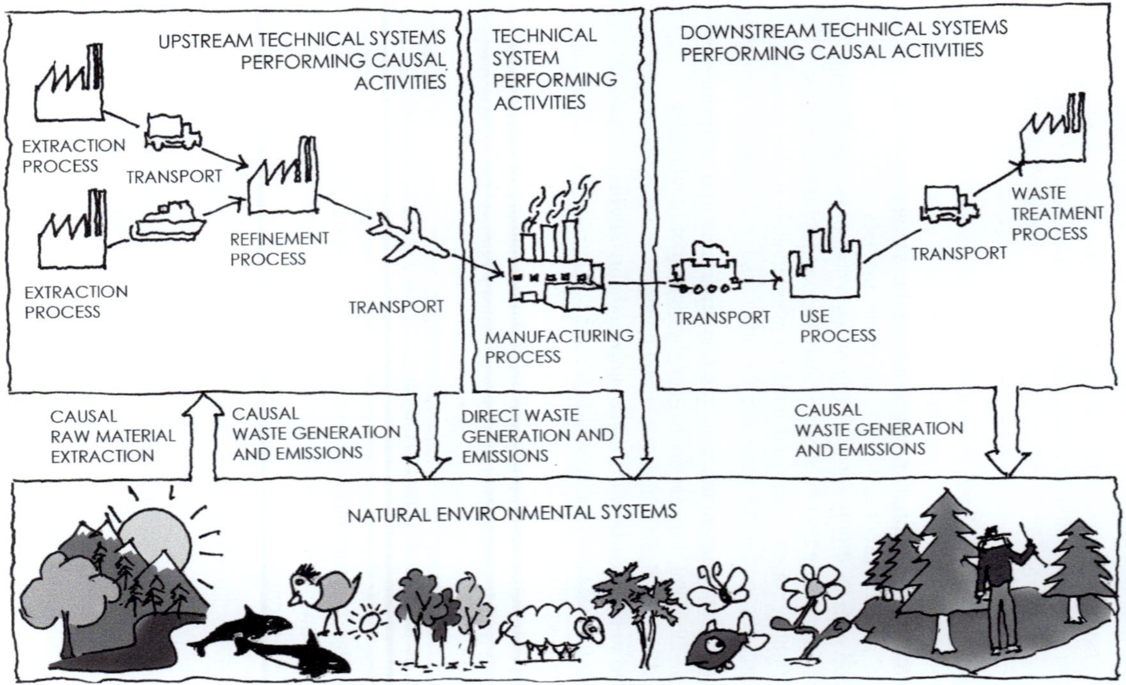

Figure 3.6 A Graphic Representation of Life Cycle Analysis

Source: Reproduced with the permission of the American Institute of Architects, 1735 New York Avenue, Washington, D.C., 20006

buildings that are beautiful are loved by the community and, as a result, are renovated and cared for.[12] Beauty and transcendence are foundational to architecture and joy in human existence and should, therefore, also be considered in our decisions. The interior of the Bird's Nest in Beijing, China, is an excellent example of beauty and form combined with strength and resource conservation, because of its structural components.

The field of evidence-based design and the Wood Councils in the US and Canada are doing research into **materiality** and the resulting neurological, physiological, and psychological reactions. The presence of exposed wood in the built environment has been shown to lower blood pressure, lower perception of pain, and increase concentration. These are just a few documented benefits of exposed wood in interior spaces. Tests also showed that the material composition of structural elements has the potential to affect occupants in the same way.[13]

3.5 BUILDING AND STRUCTURAL ADAPTABILITY AND RESILIENCE

The next step towards authentic sustainability is to change our thinking about the useful life span of a building being 25–50 years, the current norm. History tells us that it is possible to design buildings that will be structurally sound for hundreds or thousands of years. So often, building design is predicated on the current use, which is the most flexible and fluid part of the equation. We need to ask how structural elements can facilitate alternate building uses, 50, 100, and 250 years in the future.

3.5.1 Adaptability

The configuration and load-bearing elements of a structure, along with the plumbing, are the most expensive and difficult to change. These are also some of the longest-lasting building elements.

For buildings to be **adaptable** means they can be easily modified for a new use.

Because of the paramount importance of structural components, it is vital that we examine our process and ask a series of questions before we automatically design based on previous models. As Kevin Lynch, a city theorist, says, "Our most important responsibility to the future is not to coerce it, but to attend to it."[14]

Before the building's immediate use is considered, we should ask ourselves:

- What are the existing climate, **microclimate**, site, and weather forces that will be exerted on the building? Could these affect the material and design in any fundamental way?
- How could the structural components be designed to withstand or accommodate future dramatic changes in weather patterns, sea levels, or unforeseen world events?
- What are the traditional and/or vernacular building methods, materiality, and massing? How can these inform our choices?
- How can my sizing, materiality, and placement be the best for the current owner, and yet allow flexibility for future changes in use and occupancy loads?

3.5.2 Preparing for Future Changes

Structural components need to allow for, not only the loads, but also how they can be flexible and robust enough to accommodate future changes in purpose, use, surrounding density, and climatic changes.

Stewart Brand breaks a building into six broad categories that he calls **shearing layers**,[15] summarized below, to show that a building is in a state of constant change. Understanding these layers gives insight into how to create a resilient and adaptable building:

- **Site**—the geographic setting and the legally defined lot. This element is eternal.
- **Structure**—the foundation and load-bearing elements. These *are* the building and have a life of between 30 and 300 years or more.
- **Skin**—exterior surfaces. These *now* change approximately every 20 years because of fashion and technology.
- **Services**—the working guts of a building, including communication, electrical and technical wiring, plumbing, HVAC systems, and elevators. These wear out or reach obsolescence every 7–15 years. Buildings have been demolished early simply because these systems are too deeply embedded to be replaced easily.
- **Space plan**—the interior layout, which encompasses non-load-bearing walls, floors, ceilings, and finishes. Commercial spaces can change every 3 years, whereas a residence may only change every 30 years.
- **Stuff**—furniture, equipment, and personal items. This layer can change monthly or daily.

Structure is one of the most enduring layers of a building, and its design will affect the building's ability to change, ultimately affecting its preservation and sustainability for the rest of its life. As Frank Lloyd Wright said, "The sins of the architect are permanent sins."[16]

Use of a structural grid system is one way to increase the adaptability and **resilience** of a building. It does not have to be **orthogonal**, as seen below in the organic form of the Real Goods Market Solar Living Center, designed by Sim Van der Ryn (Figure 3.7). Even with the organic form, the structure has rhythm, purpose, and flexibility for future modifications (Figure 3.8). The structure is an independent element, allowing the occupants freedom of use.

SOM Architects did extensive studies of London to address the increasing need for sustainable, flexible housing, involving multiple housing types, without using **greenbelt land** or creating more **urban sprawl**. To accommodate the varied housing needs and create future flexibility, SOM came up with the concept of a **module**. The general overall

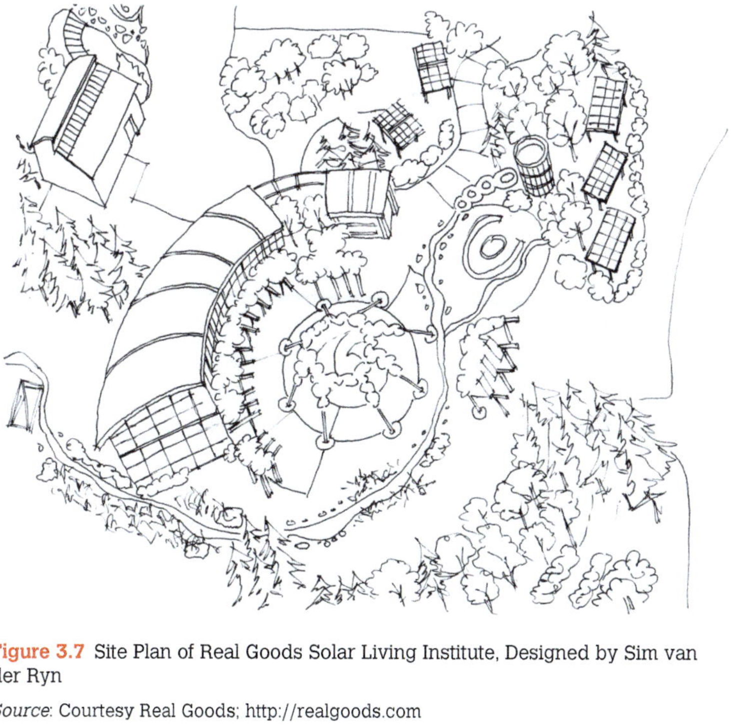

Figure 3.7 Site Plan of Real Goods Solar Living Institute, Designed by Sim van der Ryn

Source: Courtesy Real Goods; http://realgoods.com

Figure 3.8 Interior of Retail Store at Real Goods Solar Living Center

Source: Courtesy Real Goods; http://realgoods.com

grid (Figure 3.9) allows the building to be broken down into modules of varying sizes and volumes. These volumes allow for a number of shared amenities or functions. SOM found this strategy "allow[s] the building to come alive and help its inhabitants to develop a strong sense of community."[17]

As shown in Figures 3.10 and 3.11, a centralized core was designed for all the "working parts" of the building. Because of the central location, large chase, and conduits, all the mechanicals, plumbing, electricity, and technology are easily accessible.

Beauty, meaning, and cultural connections are further vital considerations for structural systems. Do our choices reinforce or deny the building's purpose, history, or present culture? Thomas Jefferson, in his design for the dormitories at the University of Virginia, used a very thin serpentine form to make a self-supporting garden wall. This created a striking visual structure that was strong and saved bricks (Figure 3.12).

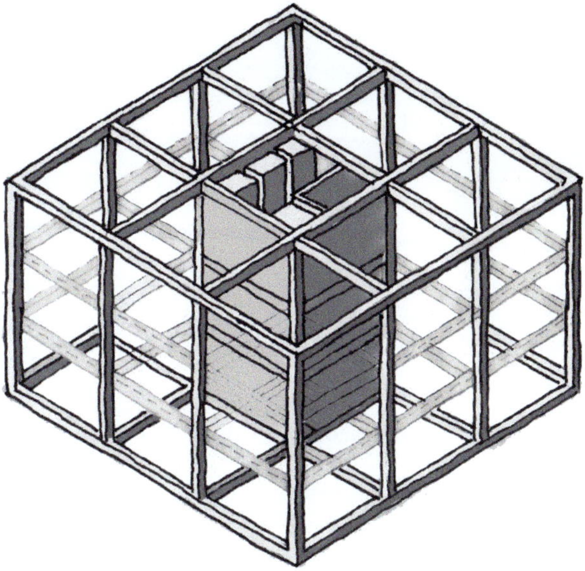

Figure 3.9 3D Illustration of Structural Grid with Central Mechanical Core

Source: Courtesy Skidmore Owings & Merrill

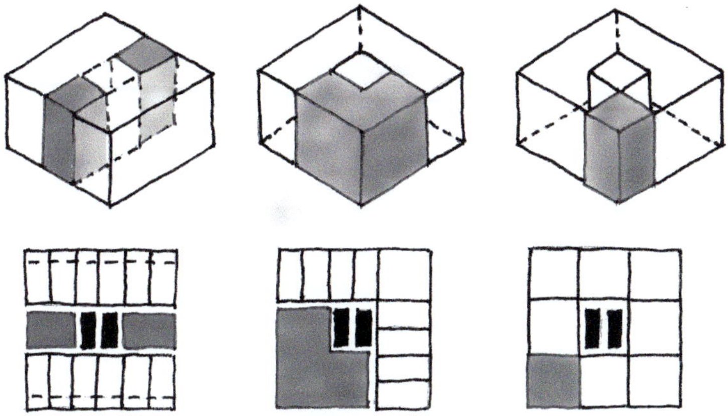

Figure 3.10 Plan and Isometric of the Varied Integration of Community and Residential Spaces Facilitated by the Structural Grid

Source: Courtesy Skidmore Owings & Merrill

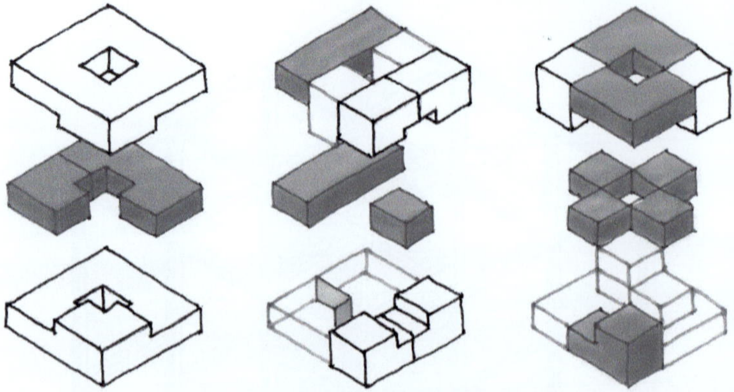

Figure 3.11 Isometric of Flexible and Integrated Use Configurations Possible with Structural Grid

Source: Courtesy Skidmore Owings & Merrill

Studies in medical facilities have shown that the same neurologic and psychological benefits that result from exposure to wood and natural views are evident when wood is used as an exposed structural element.[18] This shows that our choices have a deep impact on the occupant.

In the initial stages of design, we need to change our thinking to understand that, "Future preservation means that the building is not only built to last, but . . . has the freedom to adjust and even to change direction entirely is preserved."[19] This means that, instead of thinking of a building as a finished, completed, or static object, our aim should be towards flexibility or fluidity. The structural components of a building should be considered as a living, breathing, resilient platform to facilitate an ever-changing set of human activities.

This is a long-life, loose-fit approach. The goal is to think about a specific site with very specific design responses, and an expanded, holistic plan to best serve the present and future generations, or the "long now." This is the cornerstone of **authentic sustainable design**. Structures designed this way offer greater intrinsic value, decrease the production of new building materials, and divert vast amounts of debris from landfills.

The structural system is the most enduring part of any building and, with a little forethought, can serve its occupants and community for 50–100 or even 500 years of use.

Figure 3.12 Serpentine Garden Wall at the University of Virginia, Designed by Thomas Jefferson

3.6 EMERGING SUSTAINABLE MATERIALS AND STRATEGIES CASE STUDIES

New materials are now being used as high-performance building solutions. These emerging materials provide a great opportunity for sustainable structures because of their enhanced physical properties. We discuss some examples below.

3.6.1 Fiber-Reinforced Composites

Technically known as fiber-reinforced polymers (FRPs), these consist of two components: a reinforcement fiber and a polymer binder. They are engineered to specific performance characteristics. The advantages are durability, light weight, corrosion resistance, high strength, and low maintenance requirements. A well-known example would be the Epcot Center at Disney World in Florida.

3.6.2 PTFE Fabrics

Polytetrafluoroethylene (**PTFE**) is a Teflon coat applied to woven fiberglass fabric membrane. This combination produces an extremely durable, thermally stable surface that is light and weather-resistant. It reflects 73 percent of the solar energy, while allowing 13 percent of daylight to be transmitted. A well-known example is Denver Airport in Colorado.

3.6.3 Carbon Fiber Honeycomb

Carbon fiber reinforced polymer honeycomb mimics the properties of **balsa wood** for strength. Its chemical makeup enables 3D extrusion processes to control the alignment of fibers within the honeycomb structure, optimizing structural strength. This technology could also have implications for conductive composites as well. One of the immediate uses for this technology is the expanding size of wind turbine blades, now reaching 246 ft (75 meters) long.

3.6.4 Translucent Concrete

Translucent concrete combines the strength of concrete with the **light transmission** properties of **optical fiber**. Developed in 2001 by Hungarian architect Aron Losonzi, the mixture of fine concrete with approximately 5 percent optic fibers allows the finish material to transmit light and weigh less.

3.6.5 Cross Laminated Timber

Cross laminated timber is made of alternating layers of wood, **laminated** with non-VOC binder or a wooden peg system to create pre-manufactured building panels or structural beams. This reduces construction time, allows sculpture forms, and has superior strength. The timber also has noise-canceling and insulation properties.

3.6.6 Organic-Form Structural Systems

New design and modeling software and available building materials cause a return to organic form. They have given us the ability to build what we could previously only imagine, in a cost-effective way. The Birds Nest designed for the Beijing Olympics, shown in Figure 3.13, was created as a building of outstanding beauty, through the employment of an organic structural system of superior strength, while it used less material than a conventional grid.

3.6.7 Biomimicry

Biomimicry is the imitation of the models, systems, and elements of nature for the purpose of solving complex human problems. It looks deeper than pure **aesthetics** or form and analyzes processes, abstracts them, and then applies that information to solving real-life problems.[20] The invention of Velcro™ is a classic example of biomimicry.

Figure 3.13 The Bird's Nest, Beijing National Stadium, Olympic Green, Beijing, China

Source: Based on work by Peter23—Own work, CC BY-SA 3.0

The Eastgate building is an office complex that was designed based on study of the self-cooling mounds of the *Macrotermes michaelseni* termites. The building uses 90 percent less energy for ventilation than conventional buildings its size and has already saved $3.5 million in air-conditioning costs.[21]

3.7 CONCLUSION

Technological advances within the architectural profession are changing daily, moving us closer towards truly *sustainable* buildings. Adopting the long view of the structures we design and materials we use allows us to expand our view of their long-term effects and possibilities. John Ruskin, a writer, social thinker, and professor at Oxford University in the late 1800s said it best:

> When we build, let us think that we build forever. Let it not be for present delight nor for present use alone. Let it be such work that our descendants will thank us for . . . and that men [or women] will say, as they look upon the labor and wrought substance of them, "See! This our fathers did for us."[22]

We need to view the buildings and the structures we are designing today as a gift to future generations and the world at large. Our current society will be judged by the structures and world that we leave for those who follow us. The choice is ours to carefully and holistically analyze the multiple options for every aspect of the buildings that we design, to create a regenerative, truly sustainable structure with the potential to last hundreds of years—beautifully enhancing the community it serves.

NOTES

1. "Why the Building Sector?" Architecture 2030. Available at: http://architecture2030.org/buildings_problem_why/ (accessed February 20, 2017).
2. Intergovernmental Panel on Climate Change. *Climate Change and Biodiversity*, IPCC Technical Paper V (Geneva, Switzerland: Intergovernmental Panel on Climate Change, April 2002).
3. Bruntland Commission. *Our Common Future: Report of the World Commission on Environment and Development* (Oslo: Bruntland Commission, 1987), 16.
4. The Natural Step. The Natural Step Program. Available at: www.thenaturalstep.org (accessed February 7, 2017).
5. Mark Dekay. Integral Sustainable Design: Transformative Perspectives (New York: Routledge, 2011).
6. Dekay. Integral Sustainable Design.
7. Ashraf M. Salama. "Mediterranean Visual Messages: The Conundrum of

Identity, Isms and Meaning in Contemporary Egyptian Architecture," *Archnet-IJAR International Journal of Architectural Research* 1, no. 1 (2007): 87.

8. Rob Fleming. *Design Education for a Sustainable Future* (New York: Routledge, 2011).
9. Charles Bayer. *AIA Guide to Building Life Cycle Assessment in Practice*, ed. Russell Gentry, trans. Sarabhi Joshi (Washington, D.C.: American Institute of Architects, 2010), 22.
10. Bayer. AIA Guide to Building Life Cycle Assessment in Practice, 22.
11. Lance Hosey. *The Shape of Green: Aesthetics, Ecology and Design*, 2nd ed. (Washington, D.C.: Island Press, 2012), 7.
12. Stewart Brand. *How Buildings Learn: What Happens after They're Built* (London: Penguin Books, 1995).
13. Sally Augustin and David Fell. "Wood as a Restorative Material in Healthcare Environments" (Pointe-Claire, QC: FP Innovations, 2015), 17–19.
14. Brand. *How Buildings Learn*, 185.
15. Brand. *How Buildings Learn*, 13.
16. Brand. *How Buildings Learn*, 66.
17. Skidmore, Owings & Merrill. *Intelligent Densities/Vertical Communities: A New Housing Model for London*, NLA Breakfast Talk (London, NLA London's Centre for the Built Environment, July 2015), 32.
18. Augustin and Fell. "Wood as a Restorative Material in Healthcare Environments," 21.
19. Brand. *How Buildings Learn*, 185.
20. Biomimicry Institute. "What is Biomimicry?" Available at: https://biomimicry.org/what-is-biomimicry/#.V55evvkrJhE (accessed February 7, 2017).
21. Abigail Doan. "Biometric Architecture: Green Building in Zimbabwe Modeled after Termite Mounds," *Inhabitat*, November 29, 2012, 3. Available at: http://inhabitat.com/building-modelled-on-termites-eastgate-centre-in-zimbabwe (accessed February 7, 2017).
22. John Ruskin. *The Lamp of Memory* (London: Penguin, 2008), 233.

Performance-Based Seismic Design

Chapter 4

Paul W. McMullin

Seismic design has progressed tremendously over the past two decades. It has gone from applying a percentage of the building's weight in the horizontal direction to sophisticated computer modeling that can predict a structure's response using actual earthquake records. **Performance-based design** (PBD) is rooted in more advanced seismic design approaches and allows an owner to choose **seismic hazard** and **performance levels** in order to manage risk better.

Traditional building code seismic design reduces the largest expected earthquake effect by one-third. It inherently does not consider the maximum possible risk. Additionally, we divide the equivalent seismic force by a structural system response factor of between 3 and 8, again reducing the force—essentially lowering the performance. Consequently, a code-designed structure may be substantially damaged after a sizeable earthquake.

If the structure contains important societal or **life-safety** functions—such as a hospital, fire station, or emergency operations center—loss of use after a major seismic event is not an option. PBD provides a rational approach to design for improved performance or greater seismic risk than traditional approaches. The flow of information and effort for PBD is illustrated in Figure 4.1.

PBD allows an owner to be proactive in choosing the level of **risk** and performance of their building. It is not for every project, but, when used, brings additional thought and clarity to the design.

4.1 THE BASICS

Risk consists of two parts: exposure and hazard. In performance-based seismic design, exposure relates to performance levels, and hazard is the ground-shaking intensity. By separating the components of risk, we can effectively mitigate it.

4.1.1 Performance Levels

Performance objectives are the first step in PBD. The levels range from heavily damaged but still standing to fully functional, as outlined in Table 4.1. **Collapse prevention** is the minimum performance level, with **continuous operation** as the highest. Note that continuous operation is the same as **immediate occupancy** from a structural perspective.

Choosing these levels requires input from all stakeholders. Begin by educating the users and owner on what a code-based design provides. Then move to a discussion of seismic risk. From there, discuss

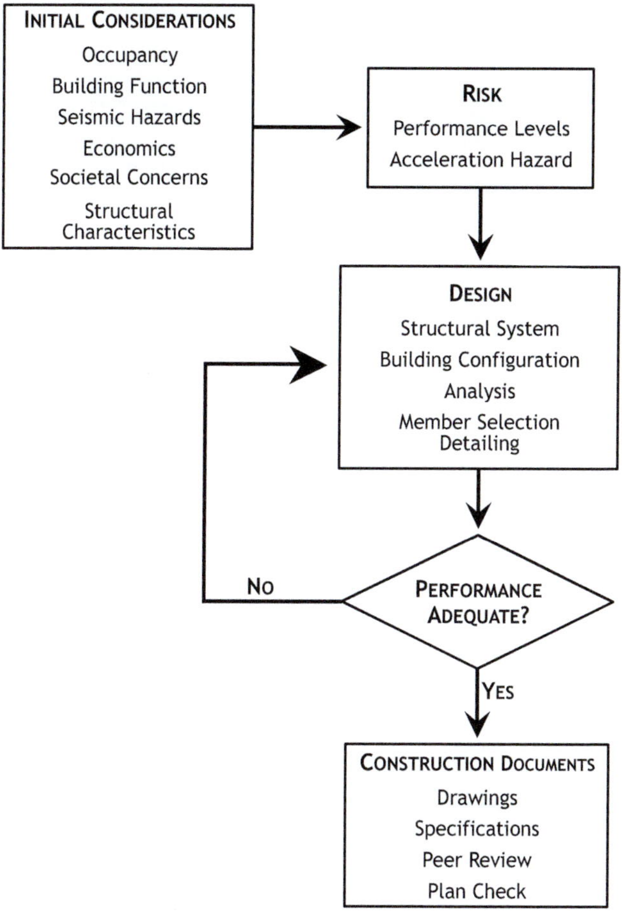

Figure 4.1 Performance-Based Design Flowchart

performance objectives and associated engineering effort and construction cost. They don't need to know all the nuances, but must understand the decisions they are making.

4.1.2 Seismic Hazard Levels

Seismic hazard levels are recurrence-based. They represent a range of seismic **accelerations** based on how often an event might occur at a

Table 4.1 **Seismic Performance Levels and Description**

	Performance Levels			
System	*Collapse Prevention*	*Life Safety*	*Immediate Occupancy*	*Operational*
Overall damage	Severe	Moderate	Light	Very Light
Lateral strength	Little	Some	Near original	Near original
Gravity strength	Barely functional	Functions well	Full function	Full function
Permanent drift	Large	Moderate	None	None
Reinforced concrete	Extensive cracking and damage	Extensive beam damage, moderate column damage	Minor hairline cracking	Minor hairline cracking
Reinforced masonry	Crushing, extensive cracking, and damage at openings	Extensive cracking throughout walls, < $\frac{1}{4}$ in (6 mm) wide	Minor cracking, < $\frac{1}{8}$ in (3 mm)	inor cracking, < $\frac{1}{8}$ in (3 mm)
Steel	Extensive beam and column connection region damage	Extensive beam damage, moderate column damage	Minor local yielding	Minor local yielding
Timber	Loose connections, extensive splitting, sheathing sheared off	Moderately loose connections, minor splitting	Minor hairline cracks in drywall	Minor hairline cracks in drywall

Table 4.1 Continued

Performance Levels				
System	*Collapse Prevention*	*Life Safety*	*Immediate Occupancy*	*Operational*
Nonstructural	Extensive damage, widespread falling hazards	Falling hazards mitigated, but widespread system failure	Generally secure, but may not function	Negligible damage
Repairable	No	Possibly	Yes	Yes
Increasing performance				

given site—the commonest being 500 and 2,500 years. Accelerations associated with these levels are available from the United States Geological Service, British Geological Survey, Japan Meteorological Agency, or similar organizations in other countries. Table 4.2 lists common seismic hazards.

More advanced analyses require seismic **acceleration time histories**. These are developed from actual earthquake acceleration records and scaled to the project seismic hazard levels, such as that shown in Figure 4.2.

Table 4.2 Common Seismic Hazard Levels

Mean Return Period (years)	*Probability of Being Exceeded*	*Description*
75	50% in 50 years	A very low level of shaking, most likely to happen
225	20% in 50 years	Low level of shaking, likely to happen
474	10% in 50 years	Moderate-to-high level of shaking, somewhat likely to happen
2,475	2% in 50 years	High-to-extreme level of shaking, unlikely to happen, often considered the maximum earthquake a fault can generate

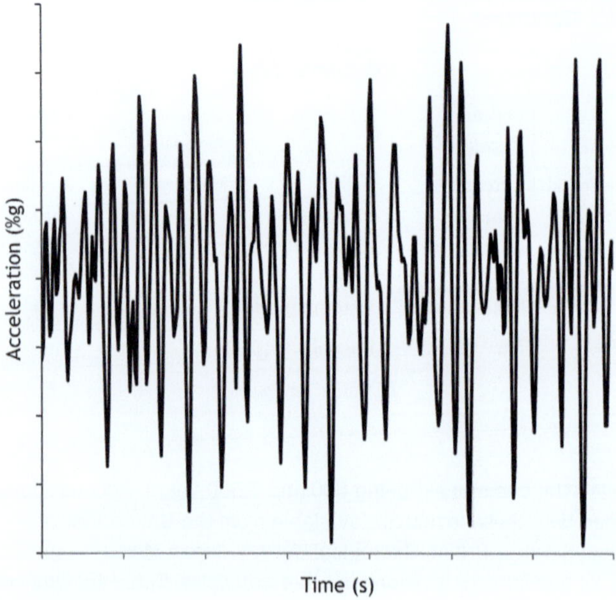

Figure 4.2 Acceleration Time History from 2011 Japan Earthquake, Tsukidate—MYG004 station

Source: Center for Engineering Strong Motion Data[1]

4.1.3 Safety Objectives

The foundation of PBD is pairing building performance with seismic ground-shaking intensity. The relationship between performance and hazard levels is shown in Table 4.3. We select two pairs: one at a higher seismic hazard with lower performance, and one at a lower hazard with higher performance. This is known as the **safety objective**, which become the basis of detailed engineering. The commonest, known as the basic safety objective, pairs collapse prevention with the 2,500-year event and life safety with the 500-year event (*k* and *p* in Table 4.3). This roughly corresponds to the code minimum. Depending on regional seismicity, one of these often drives the design criteria.

4.1.4 Design Guidance

PBD design requirements in the United States are found in *ASCE 41 Seismic Rehabilitation of Existing Buildings*.[2] It is now being used for both

Table 4.3 Seismic Performance and Hazard Levels

	Mean Return Period (years)	Probability of Being Exceeded	Performance Levels			
			Operational	*Immediate Occupancy*	*Life Safety*	*Collapse Prevention*
Earthquake hazard	75	50% in 50 years	*a*	*b*	*c*	*d*
	225	20% in 50 years	*e*	*f*	*g*	*h*
	474	10% in 50 years	*i*	*j*	*k*	*l*
	2,475	2% in 50 years	*m*	*n*	*o*	*p*

Note: For structures, Operational and Immediate Occupancy are the same
Source: After FEMA 273

new and existing structures. *ASCE 41* is a complete seismic design manual. It covers performance and hazard levels, analysis methods, and member and connector design criteria.

4.1.5 Analysis Methods

Seismic analytical methods are not unique to PBD, but are a fundamental part of it. *ASCE 41* contains four analysis methods: **linear static**, **linear dynamic**, **nonlinear static**, and **nonlinear dynamic**.

We use linear static procedures for simple structures and when a more rigorous analysis is not warranted. Static loading procedures are good for a wide range of structures and quick to implement.

Reductions in design force and structure cost can be achieved using a linear dynamic, or **modal analysis**, approach. A structure's **period** of vibration may partially insulate it from ground motion. Willowy structures usually can ride out an earthquake better than rigid structures. This approach takes advantage of structural periods of vibration, when they differ from the peak seismic acceleration spectrum. Looking at the response spectrum shown in Figure 4.3, we see the seismic acceleration is highest for fundamental periods in the 0.10–0.45-s-period range. This happens to coincide with the typical period of vibration for 1–5-story buildings and explains why these are more vulnerable than their taller counterparts. However, shorter buildings will have non-**fundamental**

periods in the longer range that contribute to the overall response. The sum of accelerations at each period gives us lower average accelerations than the static procedure.

Nonlinear static analysis, or **pushover analysis**, accounts for the yielding behavior of the structural members. To perform this analysis, we model a building frame and increase the lateral forces until a member yields. At the location of yield, we add a pin with a moment equal to the yield moment of the member. After several lateral-load-resisting members have yielded, the structure becomes unstable. Plotting displacement versus lateral force, we get a **backbone curve**, such as that in Figure 4.4. A structure that can meet its **target displacement** before its strength degrades is considered acceptable.

Nonlinear dynamic analyses take into account both the acceleration variation with time and the **yielding** of the structure, because a change in stiffness when a material yields profoundly affects the period of vibration. This is coupled with time-history acceleration data from actual earthquakes. These data are chosen to reflect the characteristics of the site and are the source of much debate. For instance, Salt Lake City has similar

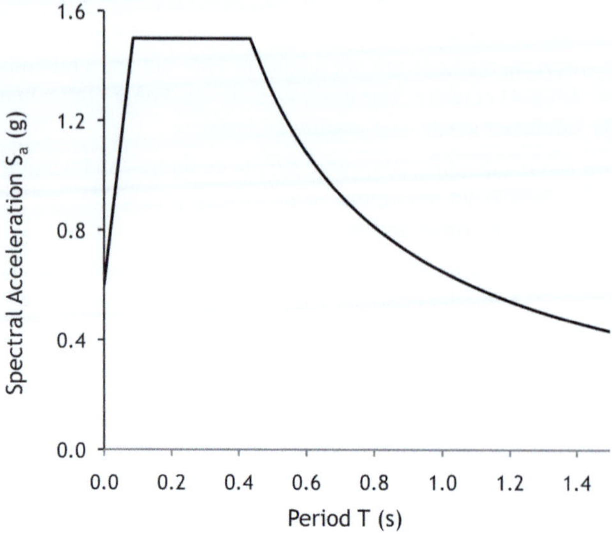

Figure 4.3 Acceleration Response Spectrum

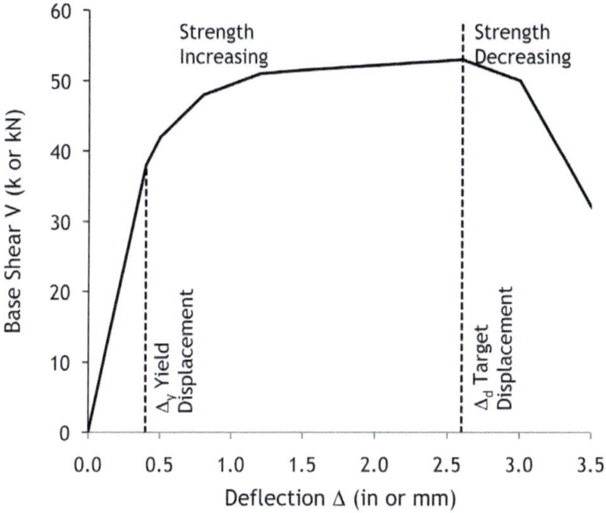

Figure 4.4 Backbone Curve for Pushover Analysis

geologic characteristics to Mexico City, and so a Mexico City earthquake dataset is often utilized for time-history data for buildings in Salt Lake City. The advantage of nonlinear dynamic analyses is that they closely model the actual behavior of the structure and may bring to light information not seen in other methods.

Regardless of analytical method, we need a rational way to validate the results. For highly complex analyses, we may use static or modal methods for validation. For simple analyses, checking lateral forces against a percentage of the building weight (10–30 percent) is common.

4.2 NONSTRUCTURAL CONSIDERATIONS

Protecting building contents and keeping lights, piping, ductwork, ceilings, and equipment from falling on people are similarly important for the functioning of buildings that have to operate after a seismic event. Seismic bracing requirements are common now in building codes, and, on critical facilities, engineers often specify life-safety equipment that will be qualified to function after a seismic event.

Seismic bracing for ceilings, sprinkler piping, and ductwork consists of light gauge strut or small-diameter cables attached every 20–80 ft

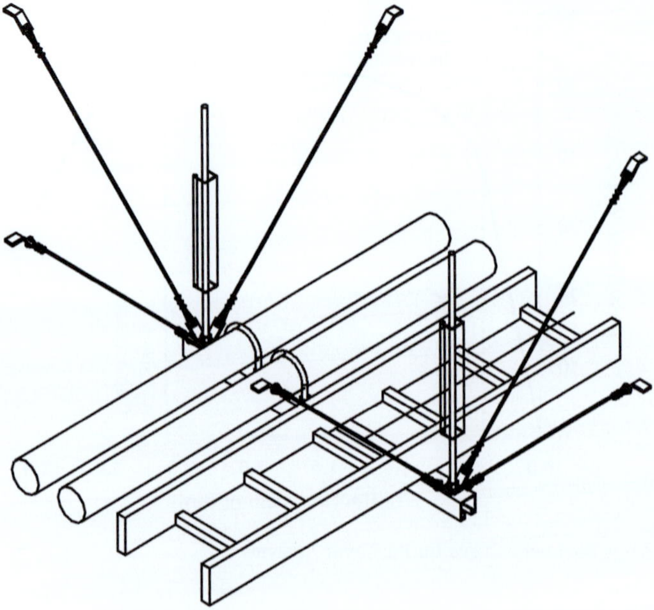

Figure 4.5 Nonstructural Seismic Bracing of Pipe and Cable Tray

Source: Courtesy Kevin S. Churilla

(6–25 m), as shown in Figure 4.5. The bracing keeps the conduit, piping, ductwork, and suspended equipment from hitting other elements, or their seismic accelerations from being amplified.

Where seismic qualification of equipment is required, we have two options: test it, or qualify it based on past experience. Few manufacturers are willing to go to the expense of seismically testing their equipment, leaving a potential void. However, databases are available that contain information on how certain types of equipment performed in seismic events. It then becomes possible to say certain types of equipment fare well in earthquakes.

Keep in mind accelerations are higher in the upper floors of a structure. So, a piece of equipment on the roof of a four-story building may see higher forces than the same piece on the ground level.

4.3 ADVANCED STRUCTURAL SYSTEMS

A number of structural systems have substantially better performance than traditional systems. Rather than solely relying on strength, they change the dynamic behavior of the structure to reduce accelerations, and therefore forces. Let's look at two such systems: **viscous dampers** and **base isolators**.

4.3.1 Viscous Dampers

Viscous dampers act like the shock absorbers in your car. As the earthquake throws the building back and forth, the dampers absorb energy, thereby reducing the force to the structure (spring). Additionally, they reduce floor accelerations, helping to protect nonstructural equipment.

Engineers place dampers in the same configuration as diagonal braces, as shown in Figure 4.6. They work in tandem with other lateral-load-resisting elements such as moment frames, increasing **damping** from 2–6 percent to 20–50 percent of critical. This significantly reduces the damage caused by the ground accelerations.

Figure 4.6 Viscous Damper Installation

Source: Photo courtesy of Taylor Devices, Inc.

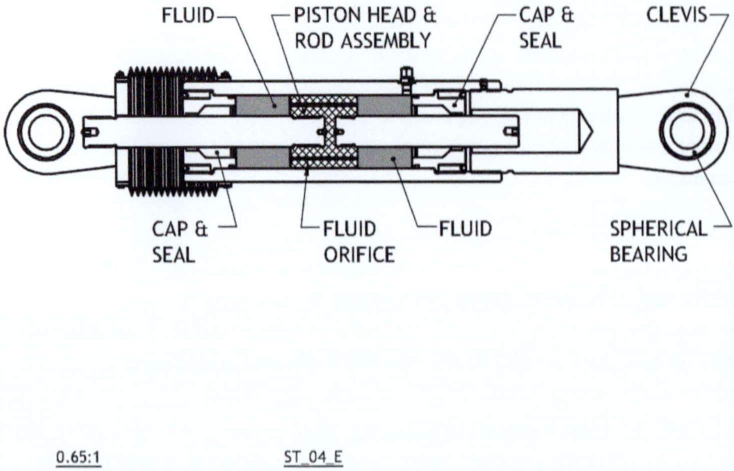

Figure 4.7 Viscous Damper Schematic
Source: Photo courtesy of Taylor Devices, Inc.

To absorb energy, the damper pushes a perforated piston through a viscous fluid, illustrated in Figure 4.7. The fluid flows with great friction, absorbing the seismic energy and slowing the structure's movement.

4.3.2 Base Isolation

Base isolation changes the fundamental period of the structure to mimic the behavior of taller structures, substantially reducing the seismic force. It is very effective on shorter structures (one to eight stories) and heavy masonry buildings. Buildings' columns and structural walls sit on isolators, which bear on concrete foundations, as shown in Figure 4.8. The isolator is made of alternating layers of rubber and steel plate, laminated together on a conical sliding bearing.

Base isolation shifts the period of the structure away from the peak shown in Figure 4.3 into the long-period range. This results in lower accelerations throughout the building, protecting structure and contents.

4.3.3 Peer Review

Because advanced structural systems are rapidly evolving and complex, *ASCE 41* requires an independent peer review. This ensures the analysis and design are thoroughly vetted. The engineers involved often have strong opinions, and the review can be fairly entertaining. Consider bringing popcorn.

4.4 WCF BUILDING CASE STUDY

When the Workers Compensation Fund of Utah (WCF) set out to build its new headquarters, shown in Figure 4.9, it wanted to maintain support for its clients after a major seismic event. The structure sits near the Wasatch Fault—considered capable of generating a magnitude 7.5 earthquake. The WCF understood that even a building designed to the current building code may not serve it adequately after a substantial earthquake.

The WCF engaged an architecture, engineering, and construction team that explored numerous options with it. This included building configuration, construction materials, structural systems, and seismic performance options. As they worked, a middle-ground performance and hazard pairing became apparent. In the end, the owner chose a basic safety objective substantially above the baseline. It consisted of immediate occupancy for a 1,000-year seismic event and life safety for a 2,500-year

Figure 4.8 Base Isolator

Source: Photo courtesy of Reaveley Associates

Figure 4.9 Workers Compensation Fund Headquarters Building, Sandy, Utah

event. Incidentally, the return period for the immediate occupancy design corresponded to WCF's standard risk interval of 1 in 1,000 years.

With the performance and hazard levels defined, the team moved forward. The final lateral system design consisted of:

- perimeter moment frames with 36 in (915 mm) wide flange columns and 21 in (535 mm) wide flange beams with SidePlate™ connections, shown in Figure 4.10; the frames and connections provided robust progressive collapse resistance—another design criterion;
- concrete shear wall core varying from 8 to 14 in (203–356 mm); the walls reduced the moment frame column weight to those available with a reasonable lead time;
- mat foundation supporting the core walls and several interior columns;
- grade beams around the perimeter to carry the moment frames.

When an owner looks at their building as being critical to their ability to stay in business, the entire discussion changes. The building goes from being an expense, or even asset, to a vital part of their operation. This opens up the ability to create a truly high-performance building that will help protect the owner's economic security. For WCF, the enhanced seismic system increased the project costs by only 3 percent—essentially the cost of new carpet. A worthwhile price to the WCF so that it could better support its clients.

4.5 FURTHER DEVELOPMENTS

4.5.1 Next Generation PBD

Performance-based seismic design originated in the 1990s, as it became apparent the prescriptive, code-based design methods could not reliably predict building performance in a seismic event. The effort generated *FEMA 273 NEHRP Guidelines for the Seismic Rehabilitation of Buildings.*[3]

Figure 4.10 SidePlate™ Moment Connection

In 2000, FEMA published a technical update and pre-standard known as *FEMA 356 Prestandard and Commentary for the Seismic Rehabilitation of Buildings*,[4] which eventually became *ASCE 41*. These documents are a strong starting point for PBD, but leaders in seismic design saw a substantial next step.

The Applied Technology Council (ATC) has been under contract with FEMA since 2001 to develop the next generation of PBD procedures. They describe the effort as follows: "The next-generation performance-based seismic design guidelines will measure building performance in terms of the potential for casualties, repair/replacement costs and downtime resulting from earthquake-induced damage to a building."[5]

Specifically, the program focuses on:

- revising the discrete performance levels that better relate to the needs of decision-makers (e.g., repair costs, casualties, interruption time);
- creating procedures for estimating repair costs, casualties, and interruption time for new and existing buildings;
- expanding nonstructural evaluation procedures to assess their damageability and post-earthquake functionality;
- refining analytical techniques to more accurately simulate building response.

In 2012, the ATC issued the results of its efforts in four companion documents, *FEMA P-58–1*,[6] *FEMA P-58–2*,[7] *FEMA P-58–3*,[8] and *FEMA P-58–4*.[9] These address the bullets above and will become the new standard in PBD in coming years. The ATC is now refining the environmental aspects and fragility curves and adding guidance for owners, architects, and engineers and it expects to be done in 2017.[10]

4.5.2 Resilient Design

In the last decade, **resiliency** has gained traction as a way to assess how a community will recover after a major event, such as a blast or an earthquake. Resilience is "the capability of a system and its components—such as a community and its essential buildings and infrastructure facilities—to quickly recover full functionality following an extreme event."[11] It provides a holistic view of the interactions of what makes a community. It focuses on robustness, redundancy, and recovery. Research is now underway to develop resiliency metrics and mitigation measures that address multiple hazards and response capabilities.

4.6 SUMMARY

PBD develops by leaps and bounds each decade and will become the basis of seismic codes in the future. It provides a rational approach that building owners can use to pair performance and hazard levels in a way that works with their risk tolerance. This allows them to spend money wisely on key structural elements, while using typical or conventional framing for the rest of the structure.

NOTES

1. Center for Engineering Strong Motion Data, CESMD. "Strong Motion Data Set." Available at: www.strongmotioncenter.org/cgi-bin/CESMD/archive.pl (accessed February 17, 2017).
2. ASCE. *Seismic Rehabilitation of Existing Buildings*, ASCE 41–13, (Reston, VA: American Society of Civil Engineers, 2013).
3. FEMA. *NEHRP Guidelines for the Seismic Rehabilitation of Buildings*, FEMA 273 (Washington, D.C.: Federal Emergency Management Agency, 1997).
4. FEMA. *NEHRP Guidelines for the Seismic Rehabilitation of Buildings*, FEMA 273 (Washington, D.C.: Federal Emergency Management Agency, 2000).
5. FEMA. *Next-Generation Performance-Based Seismic Design Guidelines*, FEMA-445 (Washington, D.C.: Federal Emergency Management Agency, 2006), 85.
6. FEMA. *Seismic Performance Assessment of Buildings*, vol. 1, *Methodology*, FEMA P-58–1 (Washington, D.C.: Federal Emergency Management Agency, 2012).
7. FEMA. *Seismic Performance Assessment of Buildings*, vol. 2, *Implementation Guide*, FEMA P-58–2 (Washington, D.C.: Federal Emergency Management Agency, 2012).
8. FEMA. *Seismic Performance Assessment of Buildings*, vol. 3, *Supporting Electronic Materials and Background Documentation*, FEMA P-58–3 (Washington, D.C.: Federal Emergency Management Agency, 2012).
9. FEMA. *Seismic Performance Assessment of Buildings*, vol. 4, *Methodology for Assessing Environmental Impacts*, FEMA P-58–4 (Washington, D.C.: Federal Emergency Management Agency, 2012).
10. Ronald O. Hamburger. Email message to author, July 2, 2016.
11. NIST. "Measures of Building Resilience and Structural Robustness Project." Available at: www.nist.gov/el/building_materials/structures/mbrsr.cfm (accessed February 20, 2017).

Structural Retrofit

Chapter 5

Jonathan S. Price

5.1 INTRODUCTION

There are many reasons to **retrofit** an existing building. These include providing a needed cultural resource, restoring a historic building, or providing intangible value to a community. Additional benefits are increased real estate value, promotion or establishment of local businesses, an increased tax base, and employment opportunities. Of equal importance, the greenest building is one that already exists.

Factors to weigh in selecting a candidate retrofit project are location, inherent value, structural condition, future use potential, and a solid funding source. We will discuss **structural integrity** in greater detail; the other aspects are beyond our text's scope.

If the building is on a historic register, the primary objective will be to preserve and restore its structural integrity, without damaging the **historic fabric**. Work on historic structures will be subject to review by the governing authority. Demolition, either through neglect or ignorance, should be opposed.

5.2 INVESTIGATING AN EXISTING BUILDING

Working on existing buildings is often challenging, more so than designing new ones. Original drawings are frequently not available, and the structure may be hidden beneath finishes. Material strengths may be difficult to determine. To field verify the original structural layout, take extensive field measurements and closely examine the structure to gauge its health. Request probes (finish removal) at strategic locations to reveal the structure, assuming drawings are not available.

Temper assumptions using your experience with similar structures from the same era. Design resources from the period provide valuable insight. If finishes or fire coatings hide the structure, then investigatory probes may be necessary, unless the alterations are nonstructural. An experienced structural engineer should determine where to remove finishes and where material samples should be taken for testing. "Do no harm" is the rule that guides our work. Removal of any structural material should not decrease the structure's strength.

5.3 CODE REQUIREMENTS

5.3.1 Uniform Building Code

The 1927 Uniform Building Code (UBC) was the first major **code** that required seismic design—mainly applicable in the western states.

Codes applicable in the east and south did not require seismic design until later, and a large number of unreinforced masonry bearing wall structures still exist throughout the US. Unfortunately, heavy buildings attract more seismic load and are more vulnerable than light wood-frame constructions.

Assuming a code was in effect at the time of construction, and if there are no significant new alterations or damage, FEMA (Federal Emergency Management Agency) offers the following engineering guidance:

> Except in certain circumstances, such as when a building is significantly renovated or altered or there is a change in its use that triggers the IBC or IEBC, the code requirements for existing buildings are those that were in effect when the structure was designed and constructed.[1]

"Altered" can include damage by fire or earthquake or other acts of God.

The structural design of most older buildings was dictated by gravity **loads**. Wind and seismic loads often were ignored, although it is hard to ignore that an older building has been standing for many years, sometimes hundreds of years, and has weathered many storms.

For areas of higher seismic activity, we consider lateral loads, even though we may not be adding mass. Older buildings are vulnerable to earthquakes, and local codes may require an upgrade, even though the building has survived many years.

5.3.2 ICC Provisions

The 2012 ICC (International Code Council) requirements related to building modifications are contained in the International Building Code (IBC)[2]—Chapter 34—and the International Existing Building Code (IEBC).[3] These codes overlap, causing some confusion as to which one applies. Our guidance to the designer is that the IBC is a stand-alone document that can be used without reference to the IEBC. The IEBC expands upon the IBC, and both codes require upgrades if the structure it is unsafe for occupancy.

The IBC has been legally adopted in every state in the US, although the edition varies between states. The IEBC has almost the same level of acceptance. Refer to the ICC website for the codes in effect in each state before starting design and understand how they will apply to your project. Communicate your code interpretations (IBC edition, live and dead load assumptions) to your client. Request their interpretation of fire rating, as it may affect or limit design options.

Seismic strengthening of older structures in areas of higher seismic activity is often required by local amendment to the code. A change of **occupancy**, resulting in a higher-risk category, triggers needed seismic upgrades.

The building official can interpret code requirements during plan reviews, and so, if there are questions about the applicable code, it is best to get them answered early, before investing a lot of design effort. Additionally, the local jurisdiction can make amendments to the code that can drastically affect your design. Be aware of these before starting.

5.3.3 Voluntary Structural Upgrades

Building codes typically allow voluntary upgrades for minor building modifications without an overall strengthening, provided that:

- the upgrades are designed in accordance with the current code;
- they do not diminish the load capacity of other structural elements; and
- they do not add other loads to the building.

5.3.4 ADA Act

The 2010 Americans with Disabilities Act (ADA) requires accessibility upgrades for major building renovations. Most older buildings are noncompliant with ADA provisions and, therefore, renovations typically include new ramps and elevators. These may require a floor area to be reframed or an area of existing framing system to be demolished for a new elevator shaft.

ADA rules apply to all public, educational, and commercial buildings—with exceptions for historic structures or "structural impracticability." Further, Paragraph 28 CFR 35.151 of the ADA requirements for public buildings states:

> Each facility or part of a facility constructed by, on behalf of, or for the use of a public entity shall be designed and constructed in such manner that the facility or part of the facility is *readily accessible* to and usable by individuals with disabilities, if the construction was commenced after January 26, 1992.[4]

5.3.5 Historic Structures

Structures on the National Register of Historical Places or a local register require greater attention to details. The Secretary of the Interior's Standards offer four distinct approaches to the treatment of historic properties—preservation, rehabilitation, restoration, and reconstruction—with guidelines for each.[5]

5.4 APPLICATION OF THE CODES

5.4.1 Compliance Options

Chapter 3 of the 2012 IEBC outlines three compliance options for existing buildings undergoing alteration.

Option 1 is a prescriptive method requiring IBC compliance for new work or damaged elements of the existing building. Exemptions are permitted for routine maintenance and minor modifications.

Option 2 branches into three alteration levels that are based on the extent of the work, Levels 1, 2, and 3.

- Level 1 modifications include those considered cosmetic or the replacement of existing materials or equipment in kind, requiring little to no structural intervention. These modifications are exempt from a full code review, assuming there is no change in use. Note that reroofing is also exempt, if the work adds less than 5 percent dead weight. However, if the building is assigned Seismic Design Category D, E, or F, then upgrades are required if the reroofing area exceeds 25 percent of the original area (ref. 2012 IEBC Section 706).
- Level 2 modifications "include the reconfiguration of space, the addition or elimination of any door or window, the reconfiguration or extension of any system, or the installation of any additional equipment" (Ref. 2012 IEBC Section 504.1). Modifications that add less than 5 percent dead load or 10 percent lateral load are exempt from a structural upgrade, assuming the existing building is not deficient with respect to the code requirements applicable at the time of construction. For building additions or if the loading threshold is exceeded, a structural upgrade will be required, and this work must comply with the current IBC (Ref. 2012 IEBC Section 807).

If an expansion joint separates an addition from the original structure, then only the addition needs to meet the current code requirements. If it is tied in both for gravity and lateral loads, it will necessitate a review of the existing structure.

- Level 3 modifications exceed 50 percent of the aggregate building area, or structural work exceeds 30 percent. Exceeding either of these limits will trigger a structural analysis of the existing building, and likely upgrades.

Option 3 is a performance compliance method described in Chapter 14 of the IEBC. This approach normally includes a structural analysis to establish the adequacy of the existing structure in support of the proposed changes.

5.4.2 Change of Use

Both the IEBC and IBC trigger a structural analysis for use changes, particularly if the new use demands a greater live load capacity. Until proven otherwise, a structure will need to be reinforced, unless it was designed for warehouse loads and the proposed use is less demanding. Let's look at some specific examples of building changes that require structural modifications.

5.5 TYPES OF MODIFICATION

5.5.1 Removal of Lateral Elements

Removal of lateral-load-resisting elements will trigger a structural analysis and potential upgrade, unless analysis shows the remaining system is adequate. Removing an interior bearing wall, cutting for exterior wall openings, and opening the roof for linear skylights all require careful thought and intervention if the net result is a weaker building. New wall openings will change load paths and reduce lateral strength and may overstress foundations at the points of redirected (or focused) loading.

5.5.2 New Elevators and Stairs

Introducing new elevator or stairway shafts requires local structural modifications. These vertical systems are often contained within masonry bearing walls, which are self-supporting. Bringing these walls up to the underside of each floor served and then cutting out the floor can be done without much difficulty. But, check that the details do not compromise fire ratings at the wall-to-floor connections. In higher seismic areas, the walls may provide needed lateral restraint.

5.5.3 Mechanical Equipment

Most building renovations include an upgraded mechanical system. Mechanical systems become outdated after 30–40 years, and very early systems only provided heat. In renovations, the design team struggles to find room for modern systems. The logical choices are in the basement, outside on grade, or on the roof.

Split mechanical systems are the least invasive structurally, because the weight of a condenser can be on grade or on the roof. The only connection between this outside unit and the suspended fan unit inside is piping. In contrast, large package-type rooftop units require dunnage or a curb and local structural reinforcing plus roof openings for ductwork. Check the added mass and wind sail area do not add more than 10 percent lateral load.

5.5.4 New Openings

Cutting holes in a structure for light transfer is more easily accommodated if they extend a full bay (to the column lines). Check the opening's effect on floor or roof diaphragms. For smaller openings, subframing is the norm, with joists or beams shored, cut, and resupported on new framing members.

5.5.5 Load Path Changes

An altered load path for lateral or vertical loads causes them to take a detour on their way to the ground, resulting in a diminishment of the building's structural integrity. Study these alterations carefully to determine if strengthening is required.

5.5.6 Vertical Additions

Vertical additions are possible if the existing structure was planned for them. Because the codes change, check the lateral system for increased loading, even if the gravity system appears to be adequate. Check both gravity and lateral-load-resisting systems, including the columns and foundations.

5.5.7 Mixed-Use Conversions

Mixed-use conversions require a fire separation between the uses (e.g., a commercial activity proposed for the ground floor, with residential above). Fire suppression "sprinklers" may be needed, with the potential of conflict between piping and existing beams.

5.5.8 Roof Insulation

Insulating roofs in areas with snow loads may result in greater snow accumulations than the framing can support. Check the roof members for the code-required snow loads when insulating the roof. Insulating an existing roof may cause moisture accumulation against the roof underside. Check the potential for condensation and take measures to avoid it.

5.6 EVALUATING EXISTING STRUCTURES

Evaluating the strength of an existing structure requires some research. What were the acceptable stresses for materials at the time of construction, and were there variations between suppliers and geographical distance that we should consider?

We know how to apply modern codes and understand the limits of modern materials. What did the original engineers or designers assume? How did they select the member sizes? To find the answers, we need to dig a bit deeper.

5.6.1 Original Strength of Materials

Research during the 1700s and 1800s laid the foundation for understanding the material limits we take for granted, although these were not written into building codes until the late 1800s and early 1900s. Code writers revise the codes on a periodic basis, responding to failures and new research.

A young engineer analyzing an old building is faced with multiple questions (some of these are repeated from above):

- How do we estimate the original material strength?
- What were the applicable design loads at the time of construction?
- What methods and assumptions did the original designer utilize?
- What do we know about potential vulnerabilities of older buildings that were not known at the time of construction?

References such as Kidder-Parker's *Architects' and Builders' Handbook*, Mark's *Mechanical Engineers' Handbook*,[6] Trautwine,[7] and the AISC website help answer the first two. Such information is shown in Tables 5.1 and 5.2.

The answer to the third question above is that they used classical, mostly determinate, analysis methods. The last question is one that an experienced engineer must answer, often with difficulty.

In some cases, we need to take samples for testing to establish material strengths. These samples are taken at locations of zero or very low stress level. This is necessary when additional loads will tax the original structure's reserve strength beyond its assumed capacity. We need to know how overstressed the structure is to determine the best method for reinforcing. In lieu of testing, we assume the steel strengths that were common at the time of construction. AISC's 14th edition specification appendix 5 lists evaluation techniques for steel structures.

Table 5.1 Allowable Stresses for Cast Iron, Wrought Iron, and Steel in 1929

Stress, Material and Shape	*Allowable Stress (lb/in²)*			
	New York	*Chicago*	*Philadelphia*	*Boston*
Compression				
Rolled steel	16,000	14,000	18,000	18,000
Cast steel	16,000	14,000		
Wrought iron		10,000		
Cast iron	16,000	10,000		
Steel pins and rivets (bearing)	24,000(s)	25,000(s)	D.S. 30,000(s)	D.S. 30,000(s)
	16,000(f)	20,000(f)	D.S. 20,000(f)	D.S. 20,000(f)
			S.S. 24,000(s)	S.S. 24,000(s)
			S.S. 16,000(f)	S.S. 16,000(f)
Tension				
Rolled steel	16,000	16,000	18,000	18,000
Cast steel	16,000	16,000		
Wrought iron		12,000		
Cast iron	3,000			
Extreme Fiber Stress-Bending				
Rolled steel beams	16,000	16,000	18,000	18,000
Rolled steel pins	20,000	25,000	27,000	27,000
Cast iron—compression side	16,000	10,000		10,000
Cast iron—tension side	3,000	3,000	3,000	4,000

Table 5.1 Continued

Stress, Material and Shape	*Allowable Stress (lb/in²)*			
	New York	*Chicago*	*Philadelphia*	*Boston*
Shear				
Rolled steel shapes		12,000	12,000	12,000
Steel web plates	10,000	10,000	12,000	12,000
Steel shop rivets and pins	12,000	12,000	13,500	13,500
Steel field rivets	8,000	10,000	10,000	10,000
Steel field bolts (unfinished)	7,000		10,000	10,000
Steel field bolts (turned and in reamed holes				13,500
Cast iron	3,000	2,000		2,000
Columns				
Steel	$16{,}000 - 70(L/R)$	$16{,}000 - 70(L/R)$ Max. = 14,000	$18{,}000/(1 + L^2/(18{,}000R^2))$ Max. = 15,000	$18{,}000/(1 + L^2/(18{,}000R^2))$ Max. = 13,500
Cast iron	$9{,}000 – 40(L/R)$	$10{,}000 – 60(L/R)$	$9{,}000 – 40(L/R)$	$9{,}000 – 40(L/R)$

Notes: L = length of column; R = radius of gyration; D.S. = double shear; S.S. = single shear; s = shop rivets; f = field rivets

Source: Mark's *Handbook*[15]

Table 5.2 Timber Allowable Stresses

Maximum stresses in timber, lbs per sq inch					
For Highway Bridges, Oo.	*Transverse loading*	*End bearing*	*Short column*	*Bearing across fibre*	*Shear along fibre*
White oak	1,400	1,300	1,000	550	300
Long leaf pine	1,600	1,300	1,000	350	200
White pine	1,100	900	700	200	150
Hemlock	950	850	650	200	100

Notes: Extreme fibre stress, in floor beams, max, yellow pine and white oak, 1,200 lbs per sq inch; white pine and spruce, 1,000, Aa, Cc.

Source: Trautwine[16]

For historic shapes, refer to *AISC Design Guide 15*: *AISC Rehabilitation and Retrofit Guide*: *A Reference for Historic Shapes and* Specifications, by Roger L. Brockenbrough, P.E.,[8] which lists, not only the strengths assumed at the time, but also the section dimensions and other engineering properties.

5.6.2 Existing Structural Capacity

It is enlightening to estimate the capacity of an existing structure based on classical methods and assumed material strengths, before utilizing modern methods. The methods are those we learn in statics and strength of materials classes.

If only a general sense of the floor or roof capacity is needed, then only a few representative members need to be assessed—assuming the building has not been modified previously, and the framing is equally spaced. Framing around proposed stairs and elevators requires reframing for the new openings and special attention because load is redirected around to openings to other members.

5.7 RETROFITTING TIMBER STRUCTURES

5.7.1 Modern Structures

Appendix A3 of the IEBC provides suggestions for retrofitting light framed structures to their foundations for enhanced seismic and wind-load resistance. Appendix A4 of the IEBC provides guidance on retrofitting timber structures having a weak first-floor structure or where the front wall is open. Appendix C1 provides guidelines for retrofitting weak gable ends of light framed structures against high wind loads (such as along hurricane oceanfront areas).

5.7.2 Historic Timber Structures

Historic timber is more difficult to evaluate, because of the wide range of species, grades, and connection types, and the potential for hidden decay or fire damage.

Decay can be hidden within the center core of heavy timber (tunnel rot) and can go unnoticed for many years, until the member fails. Methods for reinforcing are also wide ranging. If the timbers are damaged, they may need to be replaced or reinforced. If they are to be reinforced, consider how the load is transferred in and out of the new piece.

Connections may not pass current codes. Mortise and tenon joints, shown in Figure 5.1, were successfully used for many centuries. Only recently has their strength been studied, and yet load values have not been codified. When we apply modern code design rules to existing mortise and tenon connections they usually fail, but many old structures are connected with them and they have been standing for centuries. Therefore, the engineer is urged to use caution, engineering judgment, and the advice of experts, before crying wolf.

When making new connections to old timber, remember that wood does not like abrupt, heavy connections, such as large-diameter bolts. Spread

Figure 5.1 Mortise and Tenon Joinery

the connection forces out as much as is realistically possible. Do not drive lag screws into old or new timber without predrilling.

5.7.3 Fire Resistance

IBC Table 601 permits heavy timber structures in all occupancies where a fire rating of 1 hour or less is required. Heavy timber is defined by minimum sizes. For example, columns must be at least 8 in (200 mm) nominal dimension.[9]

5.8 RETROFITTING STEEL STRUCTURES

AISC provides guidance on retrofitting existing steel structures for increased gravity and lateral loads. The Steel Joist Institute and its member firms provide guidance on strengthening and modifying steel joists.

5.8.1 Weldability of Steel

ASTM A36 and A572 steels came into production after 1962 and are considered **weldable**. A carbon equivalence test tells us if the steel is weldable. In lieu of chemical analysis, a simple field test can be performed to determine if the steel on your project is weldable, as described in *Field Welding to Existing Steel Structures*.[10]

5.8.2 Beam Strengthening

If you are reinforcing a steel floor beam and deflection is a concern, relieve the dead loads before installing the reinforcing. If you are simply increasing the bending strength, the existing loads need not be relieved, because the plastic or ultimate moment capacity of the reinforced section can be realized.[11] It is best to reinforce the top and bottom flange equally, so that the neutral axis remains at mid-depth. Also, it is possible to make a non-composite beam composite by removing cores, installing shear studs through the slab, and then grouting these. This operation is tricky and should be witnessed by the inspector on a full-time basis.

5.8.3 Column Strengthening

Strengthening a column is a different matter. Assume the existing column is carrying a heavy load, but it needs to be reinforced to carry additional load. The existing stress may be close to failure, and so it is better to add area *and* moment of inertia in both the *x*- and *y*-axes. This will increase the allowable and ultimate stress levels of the existing material. Adding a steel area near the centroid is counterproductive, because the radius of gyration will decrease, which will increase the slenderness ratio.

5.9 RETROFITTING CONCRETE STRUCTURES

Concrete structures are without doubt the most difficult type of structure to retrofit, especially if original drawings are not available. Tread lightly on existing concrete members; do not increase loading without discovering the size and location of reinforcing bars.

Test specimens (core samples) sometimes can be taken through slabs to estimate the concrete's compressive strength, or other nondestructive methods can be used. Reinforcing bars may be sampled from members that are to be removed. Tensile testing should prove your assumptions regarding material strength based on age.

Changes in code provisions are issued in every new edition. Therefore, if making significant changes, verify the strength based on modern codes. Shear provisions are more restrictive, so that now, minimum shear reinforcing is required whenever calculated shear exceeds one-half the shear strength of the member. Seismic provisions require greater ductility. A hypothetical design created by graduate students at Stanford revealed that a structure designed to the 1967 UBC would be more vulnerable to collapse than the same structure designed in accordance with the 2003 IBC code.[12]

Methods of analysis, construction tolerances, and material quality control have all improved. On the negative side, shear stress limits have changed, and professional oversight during construction has diminished. Architects previously had a daily representative on-site (a clerk of the works). Today, the architect attends an on-site construction meeting once a week or twice a month at best.

5.10 INFORMATION REQUIREMENTS

For retrofit projects, add time between planning and design for discovery of building geometry, framing systems, and a structural evaluation. Allot time and fees within the design schedule for these necessary tasks. If you do not, then assume you will be discovering unknown conditions during construction, which will add cost and delays.

5.10.1 Framing Layout and Material Strength

High on the list of needed information are framing sizes and spacing. Existing floor framing plans help tremendously and will save you many hours in the early stages of design.

Initial estimates may be good enough for preliminary studies, but knowing the material strengths of beams, columns, floor decks, and so on, is important if we are adding load or modifying load pathways. Take samples from low- or zero-stress areas of the structure if needed for testing.

If plans are not available, allow enough time in your schedule (and fee) to measure bay sizes, beam spacing, floor-to-floor heights, and member dimensions. Field measurements are easier to obtain if the structure is exposed, but some older buildings are framed using concrete-encased steel or concrete beams and joists. Look for signs of distress, corrosion, rot, or settlement. These may require further study and analyses and will inform your retrofit options. Some removal of material may be necessary so that the underlying steel or reinforcing can be measured.

5.10.2 Geotechnical Information

Geotechnical and foundation condition(s) are commonly unknown. Soil borings, test pits, and a geotechnical analysis are often needed to provide design parameters such as bearing capacity. A first-pass estimate of this information is gleaned from existing footing sizes—if known—with existing loads estimated on a load rundown to determine the original design bearing pressure. This is not a substitute for engaging a geotechnical engineer, especially if additional loads are planned or if underpinning is required.

5.10.3 Other Unknowns

Abatement of hazardous materials that are discovered in older buildings falls to the owner. An environmental consultant can easily determine the presence of asbestos or lead paint. Ground contamination is more difficult to discover without environmental borings. A Phase-1 environmental assessment can alert the design team to potential contamination and the need for environment borings. Common underground contaminants include petroleum products, which can leach from deteriorated underground oil or gas tanks. This is the main reason that the EPA prohibits new underground tank installations unless they have double walls and leak detection.

5.11 EXAMPLE PROJECT

The ARCH Building on the University of Pennsylvania's campus was an underutilized three-story historic structure. Initially, it was for the campus Christian Association, but now:

The ARCH is home to three cultural resource centers: La Casa Latina, Makuu: The Black Cultural Center and the Pan-Asian American Community House. These student-focused centers share the building with the Center for Undergraduate Research and Fellowships, a new café, lounges, study spaces and a high-tech multi-use auditorium/classroom.[13]

Designed in 1927 by the firm of Thomas, Martin, & Kirkpatrick, the ARCH Building was constructed the following year by Wark Co. Builders,[14] shown during construction in Figure 5.2 and upon completion in Figure 5.3. The structure consists of a steel frame over a concrete base structure. Headroom was tight, and floor space was limited.

Although the building was structurally sound, its façade had deteriorated (Figure 5.4) owing to the elements and the effects of ivy (Figure 5.5). Additionally, the mechanical systems were outdated, and ADA access was limited.

Figure 5.2 ARCH Building, Original Construction, c. 1928

Source: Courtesy Penn Archives

Figure 5.3 ARCH Building, Completed Building, c. 1928

Source: Courtesy Penn Archives

Planned changes included a façade restoration, ADA accessibility enhancements (an elevator, new stair, and a new entranceway), refurbishing and updating a large meeting and performance space, new HVAC systems with a large rooftop AHU, and new amenities (a café, breakout and meeting spaces, lavatories).

Fortunately, copies of the original drawings were available from which we could easily determine the structural configuration, structural member sizes, and original design loads. Structural member spacing required field verification, as the drawings were approximate in this respect. Material strengths typical for the era were used, as with the design methodologies (i.e. working stress design). As a check, we used the ultimate strength design method for concrete elements.

Some information normally shown on modern structural drawings was either located on the architectural plans or not shown at all. Perhaps this

Figure 5.4 Façade Deterioration

Figure 5.5 Ivy Contributing to Deterioration

Figure 5.6 Interior, Front Staircase at Third Floor

Source: Courtesy Keast & Hood

Figure 5.7 New Entrance—Lower Level

was because details that we assume are necessary today were not necessary then, because the architect was often on-site making frequent field decisions. The renovation (Figures 5.6 and 5.7) garnered a LEED Silver certification and won the Preservation Alliance of Philadelphia's Grand Jury award in 2015.

5.12 SUMMARY

Planning, design, and construction are typical steps in new work. For retrofit projects, add sufficient time to survey, research, and analyze the existing structure. Old buildings do not surrender their secrets easily, so take time to learn them and consult with people who have experience with the types of structures you are working with.

NOTES

1. FEMA. "Building Codes." Available at: www.fema.gov/building-codes (accessed July 2016).
2. International Code Council. *2012 International Building Code* (Country Club Hills, IL: International Code Council, 2011).
3. International Code Council. *2012 International Existing Building Code* (Country Club Hills, IL: International Code Council, 2011).
4. ADA. "2010 ADA Standards for Accessible Design." Available at: www.ada.gov/regs2010/2010ADAStandards/2010ADAstandards.htm#designconstruction (accessed June 2016).
5. NPS. "The Secretary of the Interior's Standards." Available at: www.nps.gov/tps/standards.htm (accessed February 20, 2017).
6. Lionel S. Marks. *Mechanical Engineers' Handbook* (New York: McGraw-Hill, 1930).
7. John C. Trautwine. *Civil Engineers' Pocket Book* (New York: John Wiley, 1906).
8. R. L. Brockenbrough. *AISC Design Guide 15: AISC Rehabilitation and Retrofit Guide: A Reference for Historic Shapes and Specifications* (Chicago, IL: American Institute of Steel Construction, 2002).
9. American Forest & Paper Association. *Heavy Timber Construction* (Washington, D.C.: American Forest & Paper Association, 2003).
10. D.T. Ricker. "Field Welding to Existing Steel Structures," *Engineering Journal* 25, first quarter (1988): 1–16.
11. L.S. Muir. "Rehabilitation and Retrofit of Existing Steel Structures." Available at: www.larrymuir.com/Documents/Existing Structures.pdf (accessed February 20, 2017).
12. A.B. Liel, C.B. Haselton, and G.G. Deierlein. "How Have Changes in Building Code Provisions for Reinforced Concrete Frame Structures Improved Seismic Safety?" Available at: www.csuchico.edu/structural/documents/conference_workshop/Pub_Conf_2006_8NCEE_Liel.pdf (accessed February 17, 2017).
13. University of Pennsylvania. "Penn Celebrates Completion of $24.5 Million Renovation of Historic Arts, Research and Culture House." Available at: https://news.upenn.edu/news/penn-celebrates-completion-245-million-renovation-historic-arts-research-and-

culture-house (accessed February 20, 2017).

14. University of Pennsylvania Archives. "Christian Association." Available at http://hdl.library.upenn.edu/1017/d/archives/20001026025 (accessed February 20, 2017).
15. Lionel S. Marks. *Mechanical Engineers' Handbook* (New York: McGraw-Hill, 1930).
16. John C. Trautwine. *Civil Engineers' Pocket Book* (New York: John Wiley, 1906).

Geotechnical Considerations

Chapter 6

Tracy Aragon

Geotechnical design is as much art as science. Small variations in soil strata, sampling technique, and laboratory processes can all affect the accuracy of the data we use. It's the geotechnical engineer's responsibility to determine how much to trust those data, how to interpret them, and ultimately to communicate how the soil at a particular site will behave. Given the consequences, it's no wonder we use safety factors.

6.1 SOIL MECHANICS FUNDAMENTALS

The first step in any foundation design is determining what you are building on. Geotechnical conditions at your site can make or break a construction schedule and budget, and so it's important to understand them as best you can before selecting a foundation system.

6.1.1 Soil vs. Rock

What is the difference between rock and soil? Soil is a natural aggregate of mineral grains, with or without organic components, that can be separated by mechanical means. Rock is an aggregate of mineral grains bound by much stronger cohesive forces. These definitions are vague because, in many cases, there is no sharp distinction between the two—rock may weather away quickly, whereas soil may be rock hard given the right circumstances. For our purposes, we'll consider soils to be those that can be sampled and tested via the methods described in this chapter.

6.1.2 Classification Systems

Various entities have established standardized systems for describing the properties of soils, according to their ultimate objective for using soils. For example, the United States Department of Agriculture has different needs than the American Association of State Highway Officials. Each system defines clay, silt, sand, and gravel differently, according to its needs. The most common classification used for vertical construction is the Unified Soil Classification System (USCS), as adopted by the American Society for Testing and Materials International (ASTM). That's what we'll be using throughout this chapter.

The USCS uses index properties to define soil types; these are **grain size**, **grain size distribution**, **liquid limit**, and **plasticity index**. We'll define these in the following sections.

6.1.3 Grain Size and Distribution

In any soil, there will be many different particle sizes making up the whole. The simplest way of describing a soil sample is by determining these sizes and their relative proportions. This information yields indices that engineers use to describe soils. The particle sizes most used for engineering are defined by the ASTM[1] classifications, shown in Table 6.1.

If there is a wide range of grain sizes represented, then we refer to the soil as **well graded**, and, if the range is small, it is **poorly graded**. Well-graded soil provides a denser, more-compressible, less-permeable layer once compacted than a poorly graded soil—making it preferable for earth embankment dams, levees, and pavements. Poorly graded soils offer relatively high strengths, depending on other characteristics of the soil particles, and yet are more permeable, making them suitable for some types of foundation.

For coarse-grained soils—particle sizes 0.029 in (0.74 mm, or the No. 200 size) and larger—we use a set of sieves, shown in Figure 6.1, to separate the soil particles mechanically. As the sample passes through the stack of sieves of decreasing size, particles are retained on each level. We then determine the fraction of the sample that passes each level—that is, the percent of the soil particles that are smaller than each successive sieve. We plot these percentages on the semi-logarithmic particle size distribution chart, which is a useful tool for the comparison of various soil sample compositions. An example of a particle size distribution curve is shown in Figure 6.2. It is useful to note that sieves are described either by the largest particle size passing through the openings (in inches or millimeters), or by the U.S. Standard sieve number,

Table 6.1 Particle Size Limits as Defined by ASTM D422

Soil Classification	*ASTM Soil Particle Size*		*ASTM Sieve Designation*
Gravel	Larger than 0.187 in	Larger than 4.75 mm	Larger than No. 4
Coarse sand	0.0787–0.187 in	2.00–4.75 mm	No. 10 to No. 4
Medium sand	0.0165–0.0787 in	0.425–2.00 mm	No. 40 to No. 10
Fine sand	0.0029–0.0165 in	0.075–0.425 mm	No. 200 to No. 40
Fines (silt and clay)	Smaller than 0.0029 in	Smaller than 0.075 mm	Smaller than No. 200

which corresponds to the number of "wires per inch" used to make the sieve. For instance, a No. 200 sieve has 200 wires per inch in each perpendicular direction.

For fine-grained soils—particle sizes smaller than 0.0029 in (0.075 mm)—we use hydrometer analysis to determine particle size distribution. The soil sample is suspended in a solution, and a hydrometer is used to measure the specific gravity over a length of time, as the particles settle in the solution. Using **Stoke's law**, we calculate the particle sizes at each time interval and plot the resulting sizes on an extension of the particle size distribution curve.

6.1.4 Water Content

One of the most important things to know about in situ soil is **water content** (also called the **moisture content**), which is the ratio of the weight of water to the weight of dry soil solids in a sample. The water content is expressed as a percentage and is represented as *w*, or sometimes *WC*.

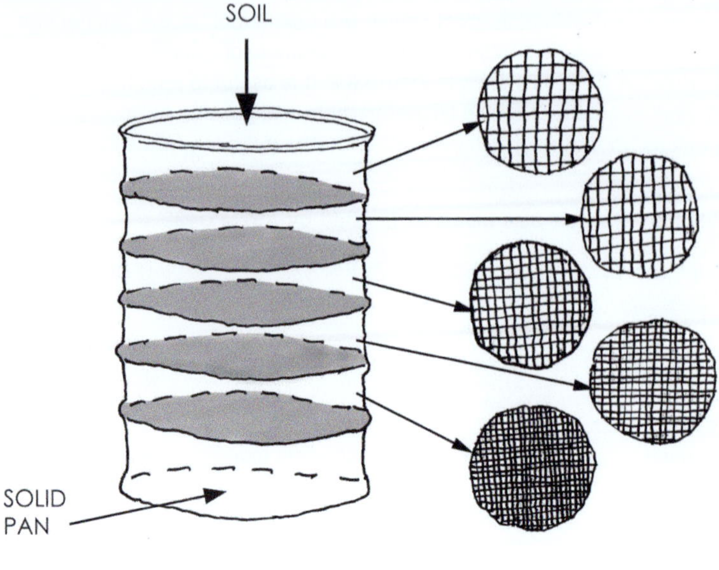

Figure 6.1 Sieve Analysis

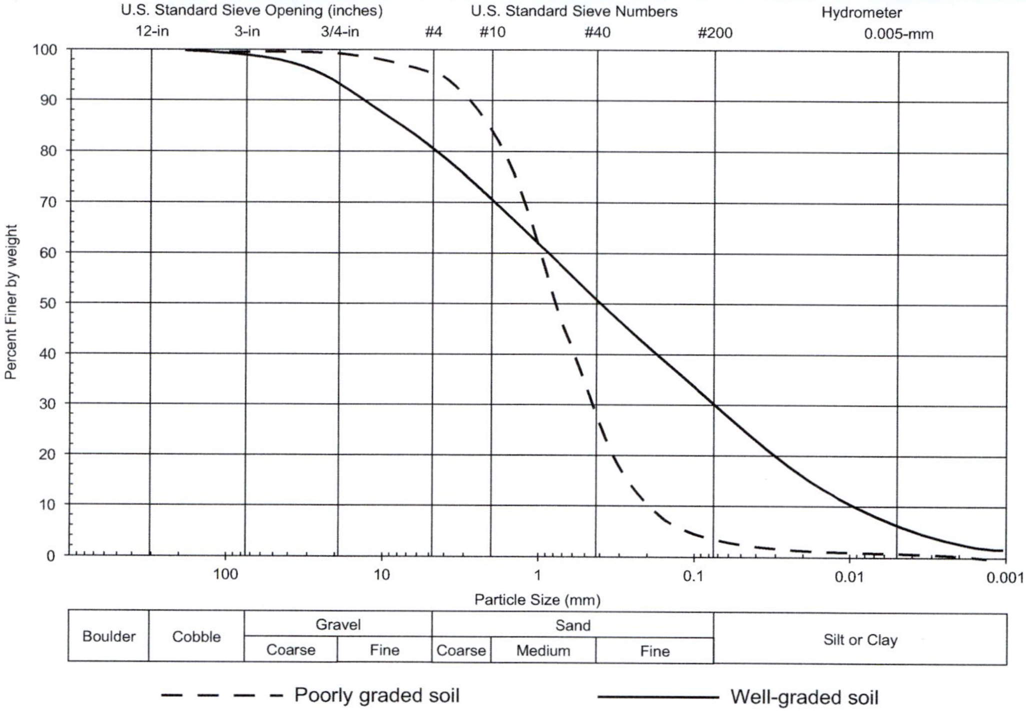

Figure 6.2 Particle Size Distribution Curve

Water contents can range from zero to more than 500 percent, but most soils have contents of less than 100 percent.

The moisture content of a soil will contribute to the density of a compacted soil layer and the way that layer behaves under loading. Imagine building a sand castle with dry sand versus wet sand. Too little water, and your castle is just a loose pile of sand; too much, and your sand slops around and doesn't hold a form. The sand never changed—just the water content. In geotechnical engineering, we watch the water content closely to ensure soils are performing the way we expect them to.

6.1.5 Atterberg Limits

Clay mineral content in soil will significantly affect its properties. It is important, therefore, that we understand how much clay is present in each sample and how it behaves. We do this is through the Atterberg tests: **liquid limit** (**LL**), **plastic limit** (**PL**), and **shrinkage limit**.

Cohesive soils, such as clays, can move from a semi-liquid state to plastic, semi-solid, or solid, depending on water content in the soil. The moisture-content levels at which these changes occur are referred to as the liquid, plastic, and shrinkage limits, or collectively, the **Atterberg limits**. The Atterberg tests are run on the portion of a soil sample that is smaller than the No. 40 sieve—about 0.016 in (0.4 mm). Let's discuss these further.

- The LL is defined as the moisture content at which a ½ in (12.5 mm) wide groove closes after 25 blows in Casagrande's liquid device (ASTM D4318).[2] This device is shown in Figure 6.3.
- The PL is defined as the moisture content at which the soil crumbles when rolled into a ⅛ in (3.175 mm) thick thread (ASTM D4318). Most of us likely conducted this test at some point on a grade school playground.
- The shrinkage limit, which is less commonly used, is defined as the moisture content at which the soil will no longer lose volume with additional moisture loss.
- The plasticity index (PI) is defined as: PI = LL – PL.

The liquid and plastic limits of a soil can be plotted on Casagrande's plasticity chart (Figure 6.4), which can show graphically what type of soil you're dealing with, depending on which zones the sample plots lie within. It's a quick way to identify the soil's type and expected behaviors.

Figure 6.3 Casagrande Liquid Limit Device

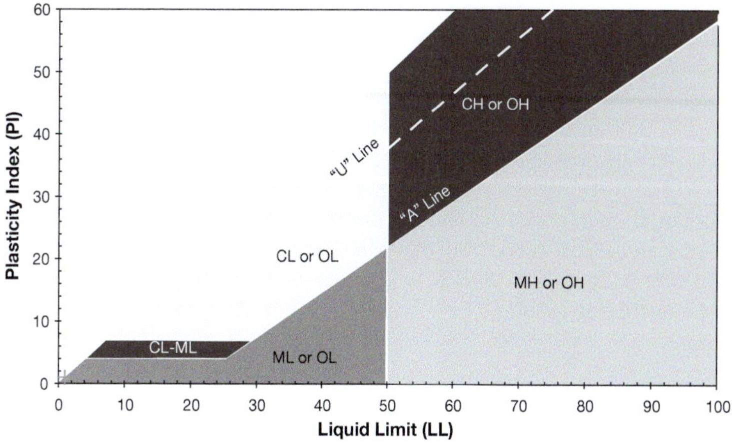

Figure 6.4 Plasticity Chart

6.1.6 Strength Parameters

As we'll discuss later when looking at bearing capacity, a common way a foundation might fail is through shear failure. The soil's shear strength is determined by (a) the loading to which the soil is subjected and (b) the soil's strength parameters, **friction angle** and **cohesion**.

The friction angle, or angle of internal friction, can be thought of simply by picturing dry sand poured out of a bucket: the slope of the pile will be uniform if you hold the bucket steady, and the sand will fall to its natural angle of repose. In a dry sandy soil, the friction angle will be the same as this natural angle of repose. If we add a bit of water to the sand, the moisture acts to increase the sand's strength, and we can form it with steeper sides, like a sand castle. This increased strength is due to the sand's increased friction angle; compaction and water content will both affect a soil sample's friction angle. Be careful though—add just a bit too much water, and the sand will start to lose its strength. It's for this reason that engineers will always specify an **optimum moisture content** for compaction during construction.

Cohesion is an electrostatic force that holds clay particles together. In high-clay-content soils, cohesive forces can play an important part in the overall strength of the soil. For other soils, such as sands and silts, cohesion is generally zero, and the strength is determined by the friction angle alone.

6.1.7 Unified Soil Classification System

Standardized classifications help engineers communicate quickly and effectively about soil properties in a field where seemingly minor variations can have significant impacts on design. Once we know the grain size distribution and the Atterberg limits, the soil sample can be classified using various flowcharts. The chart is a bit unwieldy for this book, but many charts are readily available online. There is an abundance of information in the Unified Facilities Criteria, hosted by the U.S. Department of Defense. *UFC 3–220–03FA Soils and Geology Procedures for Foundation Design of Building and Other Structures (except Hydraulic Structures)*[3] is an excellent free resource for more information about soils and foundation design.

6.2 SOIL LIMIT STATES

Our foundation design will be determined by the soil limit states, the most prominent of which are bearing capacity and settlement. We must also consider special circumstances where we have expansive or collapsible soils, frost heave, sanitary landfills, and otherwise difficult soils.

6.2.1 Settlement, Consolidation, and Compression

Settlement is one of the most important considerations in foundation design. All soils will compress or expand under loading and unloading. To understand settlement, we will need to take a look at soil composition. Soil is composed of a mixture of solid particles, water, and air; that is, the total volume of the soil can be expressed as the volume of soil, volume of water, and the volume of air. An example of this is shown in Figure 6.5.

The ratios of each of these can tell us quite a bit about a soil sample (and we can describe them in many ways), but the most common are the:

- **void ratio**—volume of voids to volume of soil;
- **porosity**—volume of voids to total volume; and
- **saturation**—volume of water to volume of voids, expressed as a percentage.

Saturation, not be confused with water content, is simply a measure of the amount of water in a soil relative to the largest amount of water it could potentially hold.

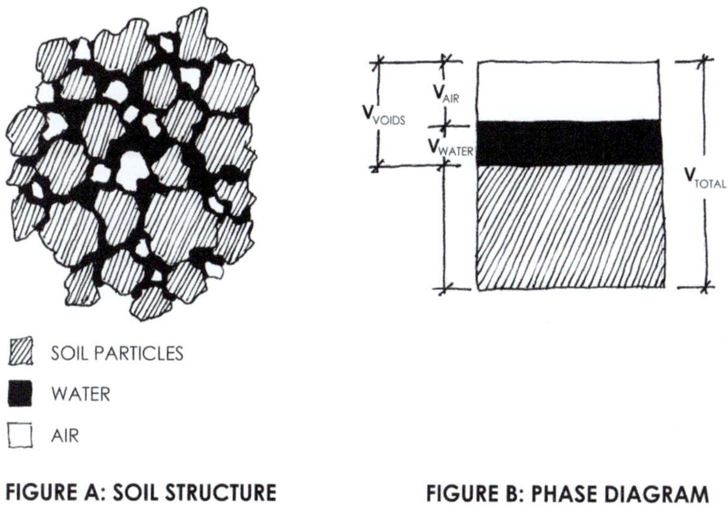

Figure 6.5 Soil Structure

There are three ways a soil will compress when loaded:[4]

1. through deformation of the soil particles themselves;
2. through compression of the air and water in the voids between the soil particles; and/or
3. through removal of air and water from the voids.

Considering the typical engineering loads: the first two are often neglected, because the pressures we are inducing are not great enough to significantly deform the soil particles themselves or the water in the voids. That leaves us with the removal of the air and water from the voids. As the air and water are squeezed out, the soil particles rearrange themselves into a denser configuration. The time it takes for this reconfiguration to occur depends on the load applied and the permeability of the soil. Settlement and shear failure are shown in Figure 6.6.

Total expected settlement is the sum of immediate settlement, consolidation, and secondary compression. The first will occur immediately upon the soil being loaded and is associated with coarse-grained soils. The second and third are time-dependent and will increase over time if the load remains on the soil. Time-dependent settlements are subject to the permeability of the soil—or how quickly the water can escape. Poorly graded sand, therefore, will have a larger immediate settlement and nearly negligent consolidation and secondary compression owing to its high permeability. Clay, on the other hand, will have lower immediate settlement and higher consolidation and secondary compression owing to its very low permeability and may, in fact, continue to settle over years and even decades. Tables 6.2 and Table 6.3 list common causes of settlement and methods for reduction and remediation, respectively.

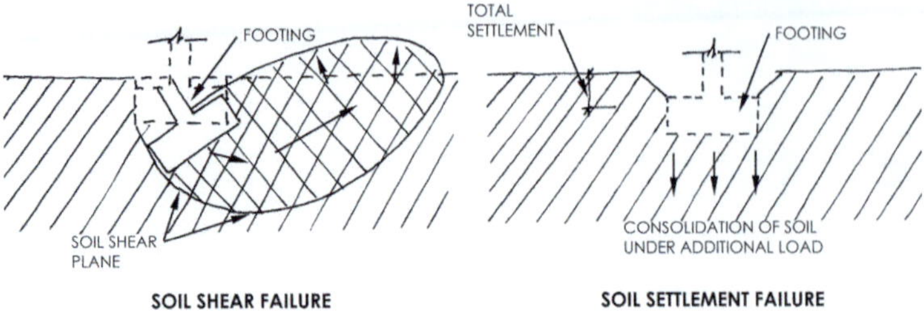

Figure 6.6 Settlement and Shear Failures

Table 6.2 Common Causes of Foundation Settlement

Cause	*Notes*
Compression of foundation soils under static loads	Soft, normally consolidated clays and peaty soils are most compressible; loose silts, sands, and gravels are also quite compressible
Compression of soft clays due to lowering groundwater table	Increased effective stress causes settlement with no increase in surface load
Compression of cohesionless soils due to vibrations	Loose sands and gravels are most susceptible; settlement can be caused by machine vibrations, earthquakes, and blasts
Compression of foundation soil due to wetting	Loose, silty sands and gravels are most susceptible; settlements can be caused by rise in groundwater table or by infiltration
Shrinkage of cohesive soils caused by drying	Highly plastic clays are most susceptible; increase in temperature under buildings containing ovens or furnaces may accelerate drying; wetting of highly plastic clays can cause swelling and heave of foundations
Loss of foundation support due to erosion	Waterfront foundations must extend below maximum erosion depth
Loss of foundation support due to excavation of adjacent ground	Most pronounced in soft, saturated clays
Loss of support due to lateral shifting of the adjacent ground	Lateral shifting may result from landslides, slow downhill creep, or movement of retaining walls
Loss of support due to formation of a sinkhole	Soils overlying cavernous limestone and broken conduits are susceptible
Loss of support due to thawing of permafrost	Permafrost should be insulated from foundation heat

Table 6.2 Continued

Cause	*Notes*
Loss of support due to partial or complete liquefaction	Loose, saturated sands are most susceptible
Downdrag on piles driven through soft clay	Loading on piles is increased by negative skin friction if soil around upper part of pile settles

Source: Technical Manual TM 5-818-1/AFM 88-3

6.2.2 Bearing Capacity

Bearing capacity is one of the most important factors for foundation design. The stress that a soil can support will depend on the soil composition, compaction, water content, depth, size, and shape of footings, and a host of other intricacies such as geologic deposition. Remember, soil behavior depends heavily on the in situ condition. A disturbed soil sample, or one transported to a laboratory, will always react to stress and strain a bit differently than it would if left in place. We can, therefore, look at bearing capacity as a rather simple concept, but know that determination of the reported values from laboratory testing is quite a bit more complex.

Ultimate bearing capacity, q_u, is the loading per unit area of foundation at which the soil experiences shear failure. At this point, excessive settlement occurs with little to no additional loading. The bearing capacity is not a function of the soil alone, and it is important to understand the various terms that engineers may use to describe it:

- **Gross ultimate bearing capacity**, q_u, is the maximum pressure possible before failure for a certain soil layer given a specific foundation size, shape, and depth. Larger foundations influence various soils at greater depths, and their capacities may be less. It is always a good idea to inform your geotechnical engineer about anticipated column loads, so that he or she can estimate footing sizes and types to tailor the subsurface investigation to your specific needs.
- **Net ultimate bearing capacity**, $q_{net(u)}$, is the total structure weight the foundation in question can support. It is the ultimate bearing capacity minus the overburden.

- **Allowable bearing capacity**, q_a, is the gross ultimate bearing capacity, divided by a factor of safety.
- **Allowable net bearing capacity**, $q_{a(net)}$, is simply the net ultimate bearing capacity divided by a factor of safety. This value is what engineers prefer to use, because it is independent of foundation depth and overburden weight.

Table 6.3 Methods for Reduction and/or Remediation of Foundation Settlement

Method	*Notes*
Use of piles, piers, or deep footings	Differential settlements between buildings and surrounding ground can cause problems
Excavate soft soil and replace with clean granular fill	Can be costly if the compressible layer is deep/extensive
Displace soft soil with weight of granular fill or by blasting	Difficult to control; pockets of entrapped soft soil can cause large differential settlements
Reduce net load by excavation	Weight of a one-story building is equal to the weight of 1 or 2 ft of soil
Surcharge or preload site before construction	Settlement is reduced by the amount which occurs before construction; preload is limited by stability considerations
Delay construction of buildings to be built on fills	Settlement which occurs before construction does not affect building; fill settlement can be accelerated using sand drains
Use a stiff foundation with deep grade beams	Can greatly reduce differential settlement
Install leveling jacks between the foundation and the structure	Building can be re-leveled periodically as foundation settles
Select a building type which has a large tolerance for differential settlement	Steel frames and metal siding can withstand large settlements and remain serviceable

Source: Technical Manual TM 5-818-1/AFM 88-3

Overburden, q_q, is the weight of the footing itself, plus any additional soil overlying the footing. Engineers usually assume, for the sake of expedited iterative calculations, that the difference in weight between the soil and the concrete is negligible, in which case overburden is the weight of the volume of soil equivalent to the volume taken up by the footing and the soil overlying it.

Now that we have defined many of the important terms, what does bearing capacity actually mean, and how do we calculate it? Referring to Figure 6.7, there are several ways in which bearing capacity failure can occur. When a soil is loaded, shear stresses develop within the soil structure. The shear failure plane occurs where these stresses are greatest. Ultimate failure occurs either when the failure plane intersects the ground surface, thereby losing all shear strength (air is not known for this quality), or when settlement continues to increase without changes in loading.

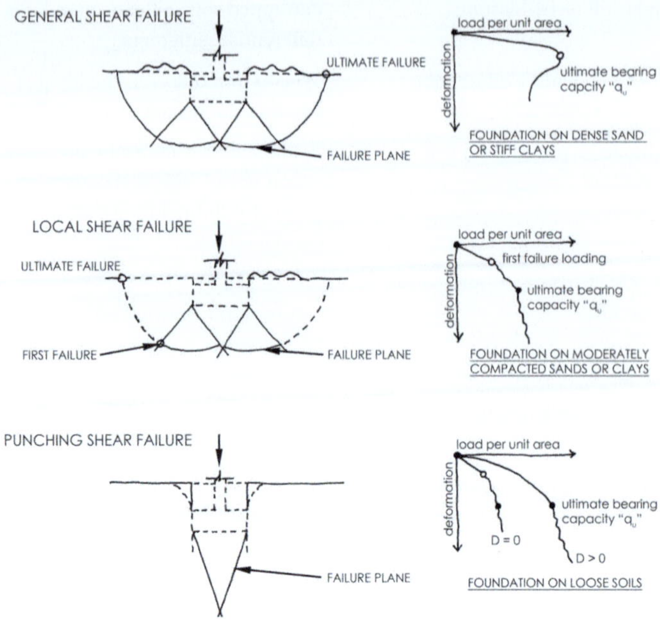

Figure 6.7 Bearing Capacity Failures

In dense or stiff soils, a foundation will settle with increased loading until it reaches ultimate bearing capacity, at which time it will continue to settle rapidly, even with decreased loading. This sudden failure is referred to as **general shear failure**.

In moderately compacted soils, a foundation will settle gradually until it reaches a first failure point. After the first failure point, the foundation settles in jerky, sudden movements until it reaches the ultimate bearing capacity, when substantial settlement will occur with minimal additional loading. This slower-moving failure is **local shear failure**.

In loose soils, a foundation will be subject to **punching failure**. Once the soil reaches the ultimate bearing capacity, the foundation will settle considerably, with little additional loading, owing to soil consolidation.[5]

Ultimate bearing capacity calculations in real-world situations are iterative. If we know our soil characteristics and the estimated structure weight, we can narrow in on the most appropriate foundation size and shape by calculating settlement and bearing capacity on a first-guess foundation design, and sizing up or down depending on the results.

Karl Terzaghi (1943) was the first to introduce a comprehensive theory for ultimate bearing capacity, and we still use his equations today. Later, George Meyerhof (1963) and Jorgen Hansen (1970) introduced adjusted factors for Terzaghi's equations that widened their applicability to include rectangular foundations, inclined loads, and additional strength provided by the soil above the bottom of the foundation.[6] As you can see below, Terzaghi's modified equation is really just a complicated-looking version of the sum of the soil cohesion, the effective stress, and the overburden weight (all times a bunch of factors to account for friction angle, shape and depth of the foundation, and the load inclination):

$$q_u = c' N_c F_{cs} F_{cd} F_{ci} + q N_q F_{qs} F_{qd} F_{qi} + \tfrac{1}{2}\gamma B N_\gamma F_{\gamma s} F_{\gamma d} F_{\gamma i}$$

$$= \text{cohesion} + \text{effective stress} + \text{overburden weight}$$

where:

c' = cohesion
q = effective stress at the bottom of the foundation
γ = unit weight of the soil
B = width of the foundation/diameter of a circular footing
N_c, N_q, N_γ are bearing capacity factors that depend on the soil friction angle φ'

$F_{cs}, F_{qs}, F_{\gamma s}$ are shape factors that depend on the length and width of the foundation

$F_{cd}, F_{qd}, F_{\gamma d}$ are depth factors that depend on the depth of the footing as well as the width

$F_{ci}, F_{qi}, F_{\gamma i}$ are load inclination factors that depend on the angle of inclination of the load

You can find tables and equations for these factors in any geotechnical textbook or online.

6.2.3 Other Concerns

Apart from settlement and bearing capacity, we may have additional concerns in what are often deemed "difficult soils." These include collapsible and expansive soils, and sanitary landfills.

Collapsible soils undergo large volume changes upon saturation. This saturation can happen unexpectedly and can result in rapid differential movement (even a plumbing leak can cause extensive structural damage in these types of soil). There are three common types of soil deposit where this may occur: Aeolian and Loess deposits, and weathered rock.[7]

Expansive soils are plastic clays that shrink and swell extensively with the subtraction and addition of water. Certain clay minerals, at a molecular level, are capable of attracting many times their volume in water, expanding the soil enough to damage foundations.

Sanitary landfills are certainly not ideal for building, but, as cities expand, there is increasing push to develop these areas. Landfills are especially susceptible to settlement, low bearing capacity, and the potential for exposure to gases such as methane. Be aware that it is not always possible to identify old landfills just by looking at a site, as the fill is covered by layers of soil.

6.3 SUBSURFACE INVESTIGATION

In a perfect world, a geotechnical engineer would have unlimited resources at his or her disposal to conduct subsurface investigations: large numbers of samples, extensive testing, and in-depth analysis. Unfortunately, we do not always have this luxury, and an engineer will be expected to design a foundation system with limited time and data. The art of soils design lies in connecting the dots using engineering judgment, based on knowledge and past experience.

6.3.1 Site Selection

Imagine giving a toddler a clear box and four cups of differently colored sand. You tell them to fill the box however they like and watch as each color of sand is poured in, some forming distinct, uneven layers, and others mixing together to form new colors.

Now, imagine that the box is your construction site (Figure 6.8), and each color of sand is a different soil with different engineering properties. You sample the soil layers in multiple locations under the footprint of the anticipated building, test the samples, and prepare your analysis based on that location. What would happen if the building location were moved? Would the samples you took still represent the soils in the new area? Perhaps . . . but perhaps not.

The only way to verify that subsurface conditions have not changed when the building location is shifted is by conducting additional sampling and testing. So, when ordering subsurface investigation, it is important to have a reasonably good idea of the location and size of the building, to avoid costly and time-consuming re-sampling at your site.

6.3.2 Methods

There are many available methods of site investigation, and each has advantages and disadvantages. The most appropriate investigation and sampling for your area likely will depend on the types of soil typically

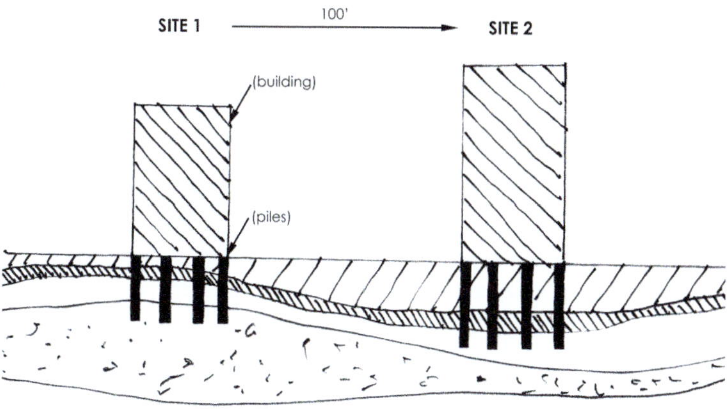

Figure 6.8 Varying Subsurface Conditions

found there and what information you really need. In many cases, several different methods will be used in a single investigation.

6.3.3 Geophysical Methods

There are a variety of investigation methods that do not retrieve samples, but instead rely on technologies such as refraction, reflection, radar, and resistivity. The information provided requires careful calibration to be accurate, and so these methods are generally used when one is looking for information about distinct interfaces, such as top-of-bedrock elevation, cavities/voids, or buried structures.

6.3.4 Reconnaissance Methods

Reconnaissance methods range from simple to relatively complex. Although there are other options out there, the following are the most common.

A test pit or trench is the simplest method and can be used for shallow investigations above the water table (or below, if you are willing to pay for dewatering). Digging deeper than 4 ft generally requires installation of sheet piles or shoring for safety reasons. Any soil samples collected from a test pit will, of course, be highly disturbed.

Auger and rotary drilling are fairly ubiquitous. In medium-to-hard cohesive soils, an auger is used to drill the hole and then removed for soil sampling at the bottom of the hole. In softer, less cohesive soils, a hollow-stem auger can be used to sample the soil through the auger bit while it is in place in the hole (the bit and stem are left in place during sampling to prevent the soil from collapsing the hole). We can take disturbed and undisturbed samples through the hollow stem, using various samplers, as well as perform standard penetration testing. An example of a drill rig is shown in Figure 6.9.

The **Standard Penetration Test** (SPT; ASTM D1586)[8] uses a split-barrel sampler to collect a disturbed soil sample and to measure the **standard penetration resistance**, or *N*-value. This *N*-value is determined by the sampler being hit multiple times with a 140-lb hammer dropped from a height of 30 in. The geologist or investigator watches the sampler and counts how many "blow counts" it takes for the sampler to drop 6 in; this is done three times (in one continuous set of hammer drops, for a total of 18 in). The blow count for the first 6 in is generally thrown out (as there may be disturbed soil in this area owing to the drilling operation), and the last two are added together to determine the *N*-value, which has as units "blows per foot." Table 6.4 correlates *N*-value to soil compactness.

Figure 6.9 Drill Rig

Source: Courtesy Jennifer Wood

Table 6.4 **In Situ Density Determination from the Standard Penetration Test**

N (blows/ft, SPT)	*Compactness of Cohesionless Soil*
< 4	Very loose
4–10	Loose
10–30	Medium dense
30–50	Dense
> 50	Very dense

A split-spoon sampler, used for collecting samples and measuring standard penetration resistance, is shown in Figure 6.10.

The **Cone Penetrometer Test** (CPT; ASTM D5778)[9] measures the resistivity when a cone tip is driven into the soil; the resistance is measured as the cone moves through the various soil strata, and soil properties and profiles are calculated. As this type of test doesn't retrieve

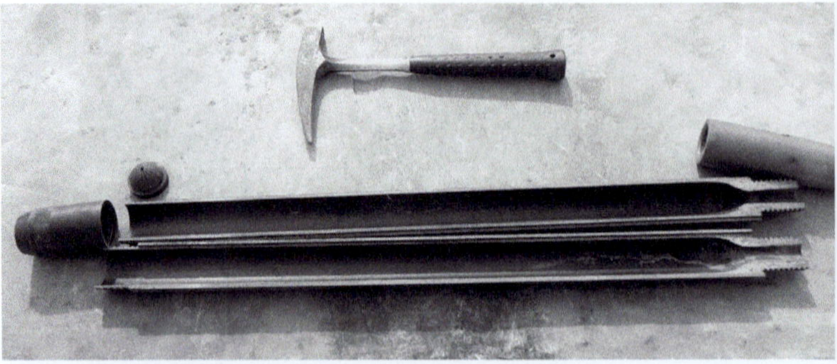

Figure 6.10 Split-Spoon Sampler

Source: Courtesy Michelle Mann

soil samples, it's a good idea to perform some additional auger drilling in several adjacent locations to verify and calibrate the data obtained from the CPT.

6.3.5 Disturbed and Undisturbed Samples

An in-place (in situ) soil will behave very differently than the same soil removed from its environment and remolded in a laboratory. For this reason, many strength, shear, and permeability laboratory tests require "undisturbed" samples; these samples can be obtained from a number of different samplers that have been developed. The most important consideration when ordering "undisturbed" samples is ensuring that they are truly undisturbed. That may mean considering things such as: Are the samples well sealed against moisture loss? Will the samples be packed in a cooler with padding? Is your site at the end of 10 miles of washboard roads? Is the drilling crew tossing them around like footballs in the field? Don't laugh—it's been done. These are all things you'll want to consider, if you are preparing to order undisturbed samples.

6.3.6 Laboratory Tests

There are many laboratory soil tests out there, but Table 6.5 lists the most common.

6.4 GEOTECHNICAL REPORTS

The geotechnical report should include a description of the site, an overview of the subsurface investigation, groundwater conditions, the

Table 6.5 Standard Laboratory Tests

Soil Property	*Tests Available*	*ASTM(s)*	*Notes*
Basic properties	Specific gravity	D854	Helps with material identification
	Moisture/water content	D2216	Needed for many other tests (Atterberg limits, density tests)
	Maximum density (Standard Proctor, Modified Proctor)	D698, D1557	For evaluating compaction requirements on fill; moisture–density relationships
Index properties	Gradation	C136, D422	For soil classification. Sieve analysis (for coarse-grained soils); hydrometer (for fine-grained soils)
	Atterberg Limits	D4318	Liquid limit, plastic limit, and shrinkage limit; computation of the plasticity index
Permeability	Constant Head Test	D2434	For free-draining soils
	Falling Head Test	D5856	For slow-draining soils
Strength	Triaxial shear	Various	Gives peak strength parameters, friction angle, and cohesion for all soils

Table 6.5 Continued

Soil Property	*Tests Available*	*ASTM(s)*	*Notes*
	Direct shear	D3080	Gives peak drained strength parameters
	Unconfined compression	D2166	Gives compressive strength for cohesive soils
	Vane shear	D2573	Gives undrained strengths
	California bearing ratio	D1883	For pavement design
Deformation	Consolidation test	D2435	Measures compression versus load and time in clay soils
	Expansion test	D427, D4546	Uses a consolidometer to measure swell and volume changes

results of sampling and testing, a geologic profile, foundation options, soil properties for foundation design, dewatering considerations, and special recommendations for construction. This section can be considered an outline for a comprehensive geotechnical report.

6.4.1 Site Description

The first section of the report should begin with a general site description, including location, principle topographic features in the vicinity, general and local geology at the site, and locations and descriptions of adjacent structures that may need consideration during construction.

Also in this section will be plan maps showing surface contours, location of proposed structure(s), and the locations and names of all subsurface investigation locations. It is often useful to show all of these on the same plan for referencing locations of borings relative to the structure footprint.

6.4.2 Subsurface Investigation

This section summarizes the results of the field investigation and includes graphic logs of all the soil borings. The logs contain details about the sampling procedure, such as date, names of field personnel, weather conditions, boring equipment used, sampler type, raw *N*-values for split-spoon penetration resistance, presence of unusual conditions, groundwater levels, and any additional notes made during the sampling.

One or more generalized geologic profiles of the site may be included as well; these profiles include representation of various subsurface strata, and may also show specific borings, groundwater levels, and select soil properties for each strata.

6.4.3 Groundwater Conditions

Groundwater is evaluated using vicinity observations, water levels found during boring, field pumping tests (if performed), and results from monitoring site piezometers (if used). Seasonal fluctuations need to be considered as well.

6.4.4 Laboratory Results

This section includes a general description of the laboratory tests performed, a range of test values, and, potentially, detailed test data for representative samples. The test results are summarized in tables and figures, and at a minimum include soil classifications, densities, moisture contents, grain size distribution curves, and Atterberg limits (plotted on a plasticity chart).

6.5 SOIL PROPERTIES

The geotechnical engineer will generate the values needed for foundation calculations from the boring logs, laboratory results, and knowledge of local soil conditions. This analysis will include a generalized soil description, net bearing capacity, settlement expectations, friction angle, frost depth, sulfate content, lateral earth pressures, liquefaction potential, collapse potential, and seismic parameters. Additional properties may be calculated, depending on the foundation recommendation.

6.5.1 Foundation System Recommendations

A comprehensive geotechnical report will recommend suitable foundation options for the site conditions, given the expected size, layout of the structure, and column loads.

There are two classes of foundation system: deep and shallow. Deep foundations use caps to transfer the building loads to piles or caissons. These in turn transmit the loads to subsurface soil and rock strata. Figure 6.11 shows various deep foundation systems. They are typically used in poor soils to support heavy loads.

Shallow foundations distribute loads over a large area near the surface and are effective in carrying moderate loads in relatively poor soils. They include spread, continuous, and combined footings, illustrated in Figure 6.12. Shallow foundations will require information to be obtained about the net bearing pressure, soil sliding coefficient, minimum embedment depth, foundation size, and settlement estimate.

Table 6.6 shows typical soil values for various soil types. Keep in mind that these values are generalized and will vary from site to site and sample to sample.

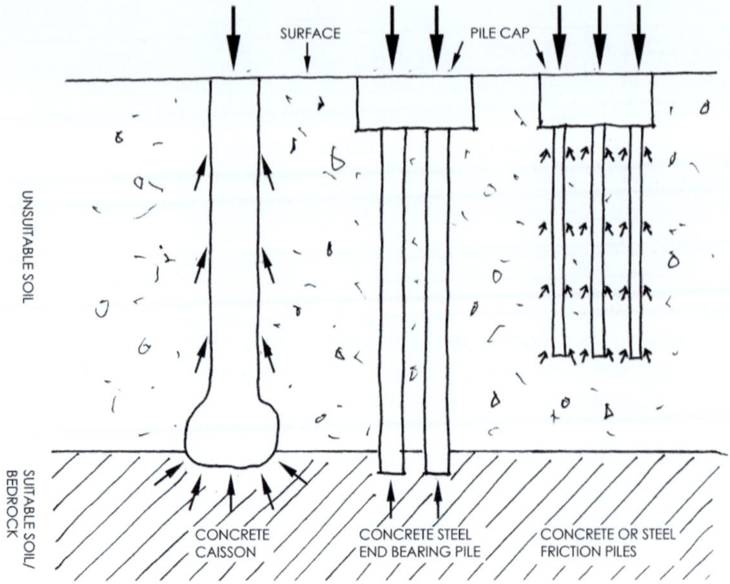

Figure 6.11 Deep Foundation Systems

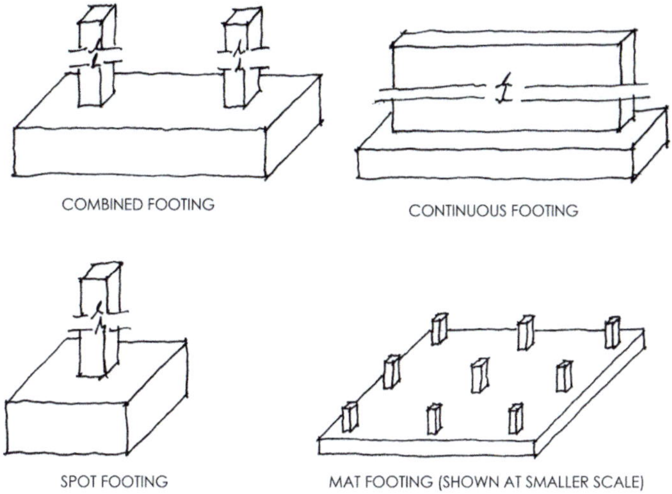

Figure 6.12 Shallow Foundation Systems

In more complex analyses, we will also need the soil modulus for the **soil spring method**[10] (also called the universal method). The soil spring method looks at the lateral soil forces on shallow pier or post foundations by modeling soil behavior as a set of springs, each exerting force on the foundation. This method is most beneficial when the soil cannot be considered homogeneous and/or when there are additional forces at play, such as a confining slab on grade abutting the foundation. Each spring can have a different stiffness to represent the variable subsurface conditions. Figure 6.13 shows an example of a simple soil spring model.

Soil stabilization is an alternative to deep foundation systems. We increase the soil strength and then use shallow foundations. Stabilization methods include rammed aggregate piers, vibratory compaction, rapid impact compaction, grouting, and preconsolidation, illustrated in Figure 6.14.

6.5.2 Construction Considerations

The geotechnical report may also contain construction requirements and additional information that will be useful for construction. This includes

Table 6.6 Typical Properties for Various Soil Types

Soil Type	*Cohesive/ Cohesionless*	*USCS Classification*	*Consistency*	*Moist Unit Weight, lbf/in²*	*Permeability Range, cm/s*	*Typical Bearing Capacity, lbs/ft³*
Low plasticity clay	Cohesive	CL	Soft	125	10^{-9}–10^{-6} nearly impervious	2,000–8,000
			Medium to stiff	130		
			Very stiff to hard	135		
High plasticity clay	Cohesive	CH	Soft	110	10^{-9}–10^{-6} nearly impervious	2,000–8,000
			Medium to stiff	115		
			Very stiff to hard	120		
Low plasticity silty/ sandy clay	Cohesive	ML	Soft	120	10^{-9}–10^{-6} nearly impervious	3,000–7,000
			Medium to stiff			
			Very stiff to hard			
High plasticity silty/ sandy clay	Cohesive	MH	Soft	105	10^{-9}–10^{-6} nearly impervious	3,000–7,000
			Medium to stiff			
			Very stiff to hard			
Silty/ clayey sand	Cohesionless	SM, SC	Loose	105	10^{-5}–10^{-2}	4,000–8,000

Table 6.6 Continued

Soil Type	*Cohesive/ Cohesionless*	*USCS Classification*	*Consistency*	*Moist Unit Weight, lbf/in²*	*Permeability Range, cm/s*	*Typical Bearing Capacity, lbs/ft³*
			Medium to dense	110		
			Very dense	115		
Sand with some gravel	Cohesionless	SW, SP	Loose	115	10^{-3}–10^{-2}	6,000–12,000
			Medium to dense	120		
			Very dense	125		
Gravel–sand mixture	Cohesionless	GW, GP	Loose	135	10^{-3}–10^{-1}	8,000–16,000
			Medium to dense			
			Very dense			

optimum water content and compaction requirements for each area of subgrade, fill, and backfill, as well as dewatering requirements, possible sources for fill and backfill, risk to (and protection of) neighboring structures, and special precautions and recommendations. If extensive dewatering will be necessary, the maximum anticipated pumping rate and flow per foot of drawdown should be included as well.

It is common for construction contractors, though generally well intentioned, to misinterpret the design intent of some geotechnical requirements, such as compaction and water content. For instance, if a report calls for 90 percent compaction of a sandy substrate, some construction personnel may take that to mean 90 percent minimum, figuring more compaction is better. In fact, the design may have been for something like a drainage layer, and a higher level of compaction may cause the feature to behave differently than intended. This is another

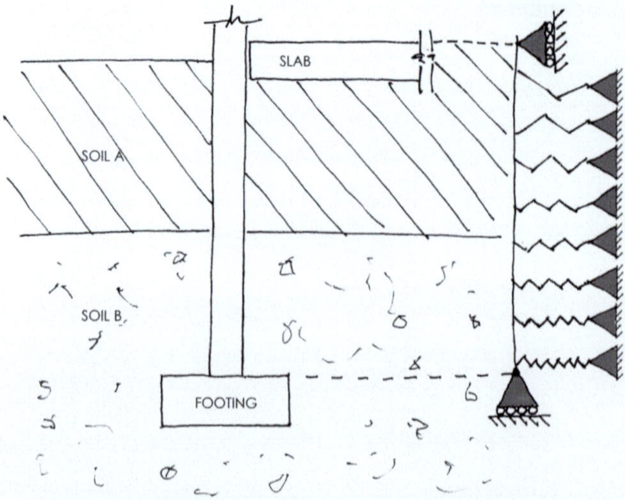

Figure 6.13 Simple Soil Springs

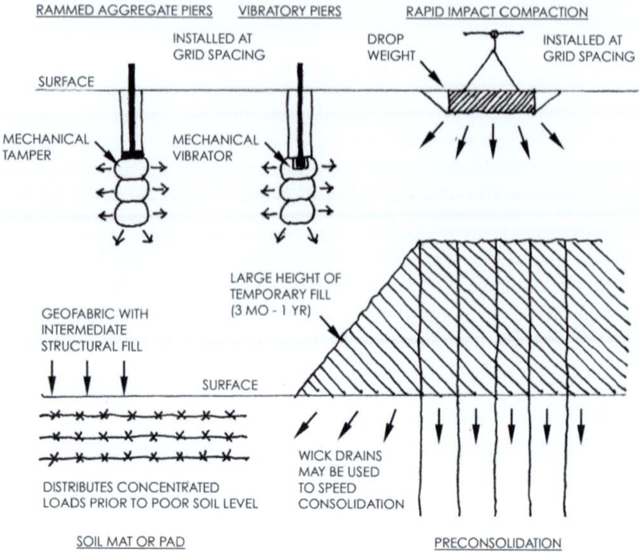

Figure 6.14 Soil Modification Systems

excellent reason to have a geotechnical engineer on-site, or at least heavily involved during construction.

6.6 OTHER CONSIDERATIONS

6.6.1 Construction near Other Structures

We do not always have the advantage of sites devoid of neighboring structures. In this case, we must consider the effects of construction. We may need to stabilize other foundations during excavation at our site, or monitor nearby structures for negative vibration effects—and all this must be coordinated with the owners of these structures. The geotechnical report should contain recommendations for protection of neighboring structures.

6.6.2 Water

High water tables, or high water pressures in lower strata, will affect foundation construction and need to be designed around or mitigated. Groundwater control methods include interception and removal (via sumps, wells, or drains), reduction of pressure, and isolation of the excavation (via cutoff walls, grout curtains, or freezing). Again, we must consider the effect on neighboring structures: lowering a water table will result in increased pressure on existing surrounding foundations. The dewatering method will depend largely on site conditions, but will also take into consideration designer and contractor capability. It is not unusual to require the contractor to provide a dewatering design for designer approval. An extensive discussion of dewatering methods and practice is available in the Army Technical Manual 5–818–5, *Dewatering and Groundwater Control.*[11]

6.6.3 Design vs. Field Conditions

The importance of continued geotechnical design into construction cannot be overstated. No matter how many holes we punch into the ground, the geotechnical design is still based on the assumption that conditions below the surface match what is seen in the soil sampling and testing. Construction excavation and other activities will confirm (or, in some unfortunate cases, contradict) what was seen during subsurface investigation. It is far more cost-effective to include the geotechnical engineer during excavation and foundation construction, where inconsistencies can be quickly identified, than to risk discovery of unexpected settlement and deformation later in the process, or even after construction is complete.

6.7 SUMMARY

There are many variables that can affect your foundation design. A thorough understanding of your site is an important first step to ensuring the stability of your structure's foundation. A robust subsurface reconnaissance program before design, and continued interaction with the geotechnical engineer during construction, will allow for fitting design and mitigate risk throughout the project.

NOTES

1. ASTM. *Standard Practice for Classification of Soils for Engineering Purposes* (Unified Soil Classification System), ASTM D2487–11 (West Conshohocken, PA: ASTM International, 2011).
2. ASTM. *Standard Test Methods for Liquid Limit, Plastic Limit, and Plasticity Index of Soils*, ASTM D4318–10e1 (West Conshohocken, PA: ASTM International, 2010).
3. U.S. Army. *UFC 3–220–03FA Soils and Geology Procedures for Foundation Design of Buildings and Other Structures*, Technical Manual TM 5–818–1/AFM 88–3 (Washington, D.C.: Joint Departments of the Army and Air Force, 1983), Chap. 7.
4. R.D. Holtz and W.D. Kovacs. *An Introduction to Geotechnical Engineering* (Englewood Cliffs, NJ: Prentice-Hall, 1981).
5. B.M. Das and S. Nagaratnam. "Settlements of Shallow Foundations on Granular Soil—An Overview," *International Journal of Geotechnical Engineering* 1 (2007): 19–29.
6. D.M. Das, *Principles of Foundation Engineering* (Independence, KY: Brooks/Cole-Thompson Learning, 2004).
7. R.B. Peck, W.E. Hanson, and T.H. Thornburn. *Foundation Engineering* (New York: John Wiley, 1974).
8. ASTM. *Standard Test Method for Standard Penetration Test (SPT) and Split-Barrel Sampling of Soils* (West Conshohocken, PA: ASTM International, 2011).
9. ASTM. *Standard Test Method for Electronic Friction Cone and Piezocone Penetration Testing of Soils*, ASTM D5778 (West Conshohocken, PA: ASTM International, 2012).
10. D.R. Bohnhoff. "Modeling Soil Behavior with Simple Springs, Part 1," *Frame Building News* (April 2014): 49–54.
11. U.S. Army. *Dewatering and Groundwater Control*, Technical Manual TM 5–818–5/AFM 88–5 (Washington, D.C.: Joint Departments of the Army and Air Force, 1983).
12. ASTM. *Standard Test Method for Particle-Size Analysis of Soils* (withdrawn 2016), ASTM D422–63(2007)e2 (West Conshohocken, PA: ASTM International, 2007).

Structural Integrity

Chapter 7

Paul W. McMullin

Figure 7.1 Bridge Crane in Industrial Application

How long should a structure last? The answer to this question is as multifaceted as the structures around us. Longevity, or lack thereof, is about **fracture**, **fatigue**, **corrosion**, concrete **deterioration**, **creep**, and **wear**. We see examples every day. The rusty bridge girder, the spalling concrete stair, or the newsflash failure caused by an unnoticed crack. **Structural integrity** is often relegated to the background, until something goes wrong. Traditional structural engineering codes do little to address the breadth and depth of considerations required to ensure long-term function of structures in moderately to severely aggressive **environments**.

Bridge and industrial structures, such as that in Figure 7.1, require much more thought about corrosion protection, fatigue, or wear than typical architectural structures. Hopefully, this chapter will open your mind to the challenges, guide you to some of the important questions you may ask, and maybe even answer a few.

7.1 IT'S COMPLICATED

Structural integrity is complicated, but that's OK. Embrace it! Oversimplification and ignoring past mistakes lead to things that matter being missed. Civil engineering history is punctuated by things we missed—and sometimes minimized—leading to unnecessary loss of life and property.[1] If you embrace the challenge, you may not have all the answers, but you will know more of the questions.

Current civil engineering fatigue design illustrates our tendency to oversimplify complex things. Why? It makes the design easier—initially. Who would not like that? However, in the world we live in, the problem and the solution are much more complex.

If we look at the way engineers talk about fatigue and fracture, we see where fantasy distorts reality. Civil codes talk about a **fatigue threshold** in steels.[2] Essentially, this is a fantastic wonderland wherein, if the stresses are low enough, fatigue will never be a problem. Thousands of bridges and industrial structures in our country prove this concept is a fantasy.

Embrace the complicated nature of structural integrity. Though the variables that influence it are numerous,[3] as shown in Figure 7.2, let them be a source of inquiry into better understanding of the challenge. To this point, Pellini, while at the Naval Research Labs, stated:

> In summary, the present trends in fracture research emphasize an ever-increasing sophistication in the treatment of the problem—building upon rather than eliminating past knowledge. The great variety of fracture research evolves from the need for attention to widely different problems which have special features. Therefore, the engineer should not expect that fracture-safe design should ultimately evolve to a single generalized procedure, but rather to a variety of procedures that overlap and integrate into a coherent pattern.[4]

More than a nice quote—it is wisdom for the ages.

7.2 FRACTURE

Fracture is the separation of two materials that once were together. Whether it is an initial crack (Figure 7.3), or a completely separated section (Figure 7.4), fracture can have catastrophic consequences. Let's spend some time understanding how cracks form and grow and how to predict their behavior.

7.2.1 Crack Growth

Consider the following four stages of crack growth: **nucleation**, short and long growth, and final instability. The first three phases are shown in Figure 7.5. The nucleation phase is the period of time required for a crack to form—assuming one does not yet exist. Next, the short crack path is highly dependent on **grain** structure and is often described as "tortuous."

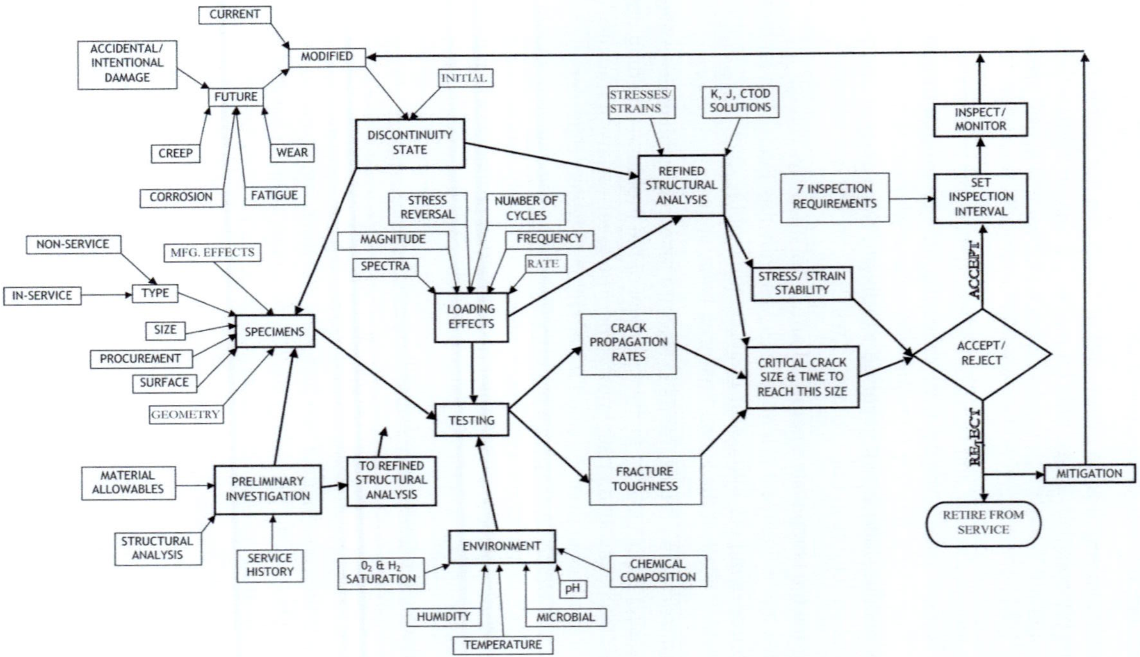

Figure 7.2 Parameters that Have Input into Holistic Structural Integrity

Figure 7.3 Surface Cracks in Truss Eyebar

Figure 7.4 Complete Fracture in a Gas Pipeline

The long crack phase is smoother, larger, and less influenced by grain, as illustrated in Figure 7.6. Final instability is often **fast fracture**, with the crack moving at the speed of sound in the material.

In addition to the phases of crack growth, there are three modes of crack loading: opening (Mode I), sliding (Mode II), and tearing (Mode III), illustrated in Figure 7.7. In Mode I, cracking is characterized by stresses and displacements normal to the crack surfaces. Mode II cracking is characterized by in-plane shearing stresses, with associated crack displacements in the plane of the crack and perpendicular to the crack leading edge. Mode III cracking is caused by out-of-plane shear, with displacements also in the plane of the crack, but parallel to the crack front.

7.2.2 Constraint

We are told by seasoned engineers to avoid over-welding joints. Why? It certainly has an economic and environmental impact, but, more importantly, not over-welding reduces **constraint**. Reduced constraint increases toughness and energy absorption.

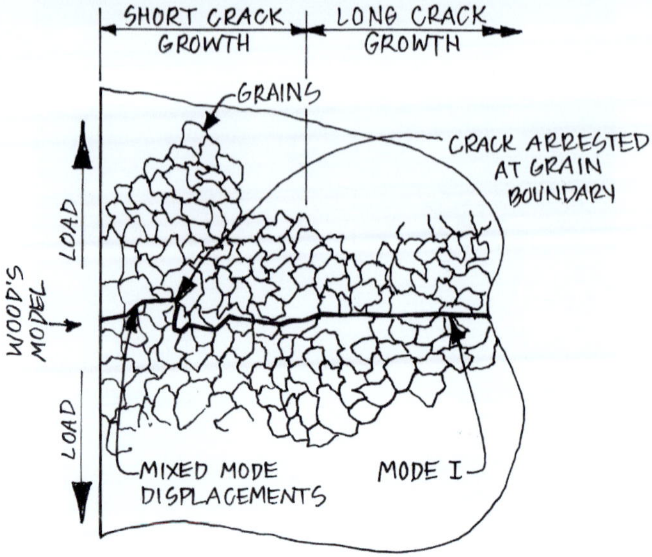

Figure 7.5 Nucleation, Short and Long Crack Growth Phases and Paths Relative to Grain Structure

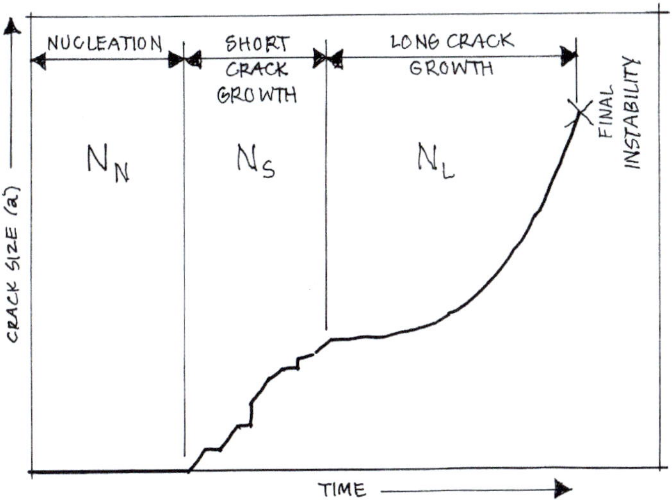

Figure 7.6 Crack Growth Phases as a Function of Size and Time

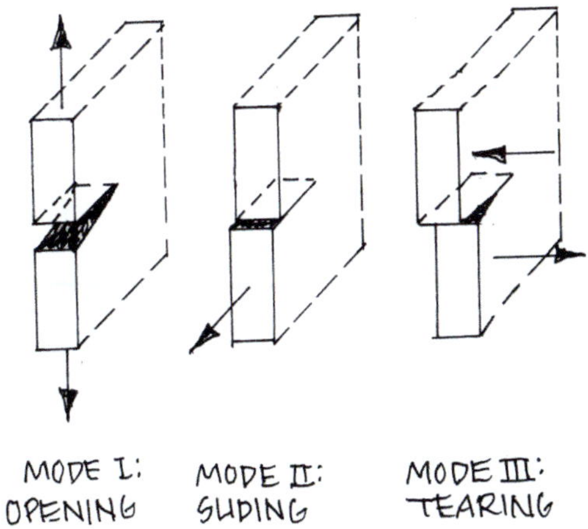

Figure 7.7 Crack Opening Modes

Constraint refers to a material's inability to deform because the surrounding material restrains it. The intersection of a beam and column flange at the webs, shown in Figure 7.8, is a common example of a constrained joint. When a material is constrained, a tensile stress is

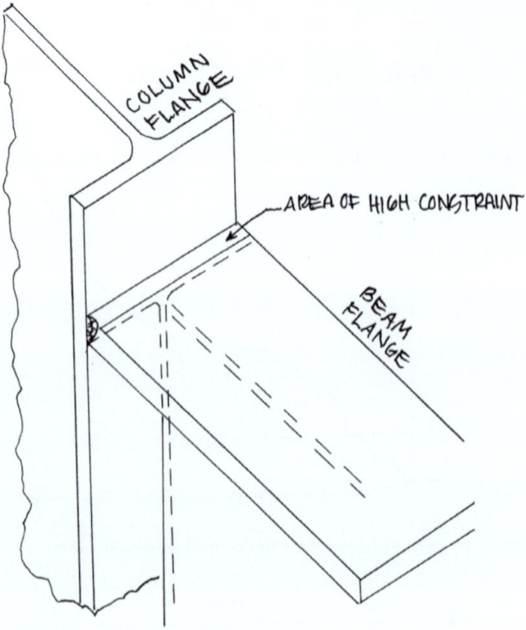

Figure 7.8 Constrained Beam-to-Column Joint

created in both directions perpendicular to the applied stress owing to Poisson's effect, as illustrated in Figure 7.9. **Triaxial stress** states reduce the principal shear stresses in materials. This in turn diminishes plastic deformation, which occurs on planes of principle shear stress owing to slip. To come full circle, the reduction in a material's ability to plastically deform reduces its toughness, or ability to absorb fracture energy.

7.2.3 Fracture Toughness

Toughness is to the boxer as strength is to the weight lifter. It is a measure of the amount of abuse a material can take. More technically, fracture toughness measures crack resistance.

Fracture toughness, like all structural properties, is influenced by size. As size increases, toughness—and strength, stiffness, and stability—decrease, as shown in Figure 7.10. As we move from a test specimen to component test to full-scale structure, there is a general trend of decreasing properties.

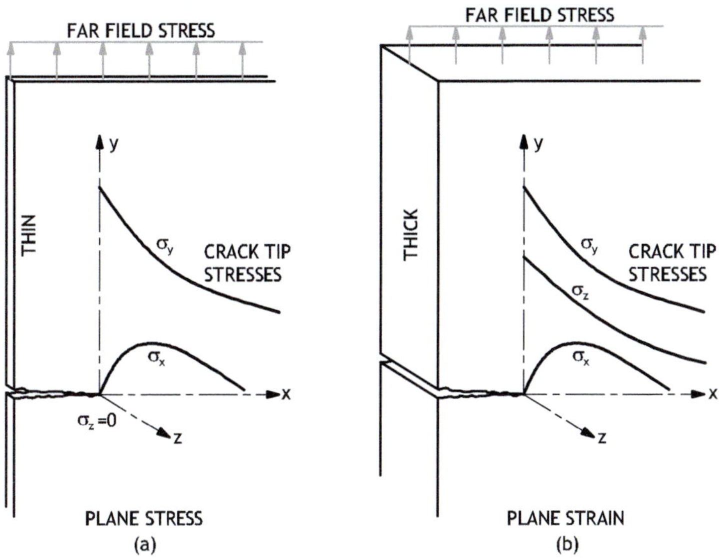

Figure 7.9 Constraint-Induced Stress State near Crack Tip

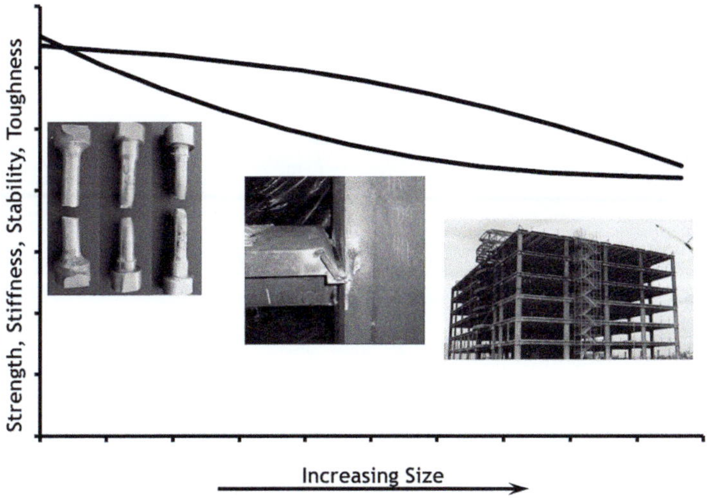

Figure 7.10 Conceptual Material Property Variation with Change in Size

Source: Photos courtesy Bill Komlos

Specific to fracture toughness, as the material gets thicker, the toughness decreases, as shown in Figure 7.11. However, after a certain point, the toughness remains the same—known as the **plane strain** fracture toughness.

Toughness is also highly influenced by temperature. Figure 7.12 shows the transition curves for a Charpy and Dynamic Tear test of the same

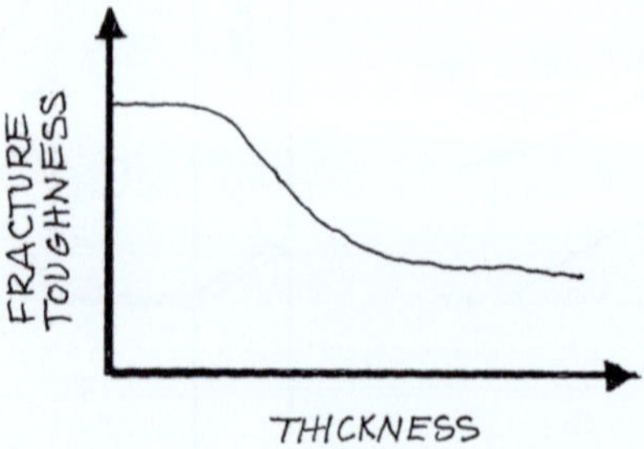

Figure 7.11 Fracture Toughness as a Function of Thickness

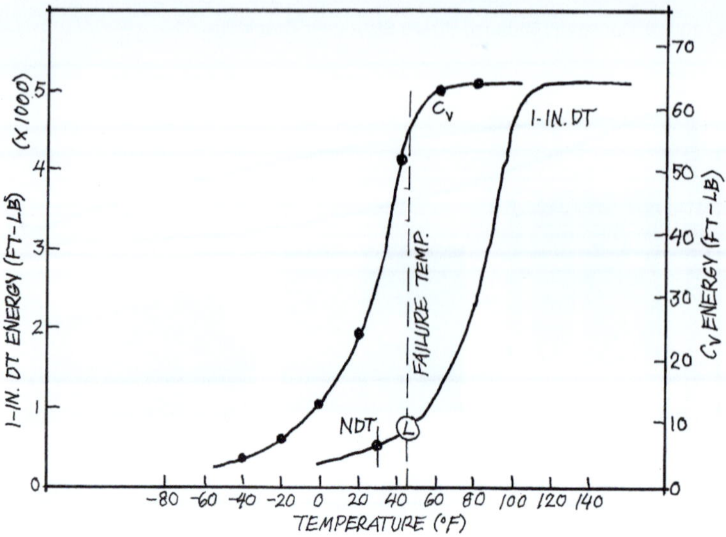

Figure 7.12 Toughness as a Function of Temperature, Specimen Size, and Test Method

Source: After Pellini[9]

material. As temperature decreases, so does fracture toughness. Note how the test type changes the location of the curve. The curve to the right is a 1 in (25 mm) thick specimen, whereas the Charpy curve is based on the standard ½ in (12.5 mm) thick specimen. If the material was 1 in (25 mm) thick, and we relied on the Charpy test, we would grossly miss the true transition temperature. This was the case in the Liberty ship failures during World War II, many of which suffered complete fractures across their hulls in port.

Finally, **strain rate** (rate of applied loading) affects fracture toughness. As the strain rate increases, the toughness decreases. It's worth considering this for impact-type loads and using tests that capture this phenomenon.

Fracture-mechanics-type tests take us from general correlations to direct analysis of crack behavior. The commonest test is plane strain, compact test specimen that determines K_{IC}, illustrated in Figure 7.13.

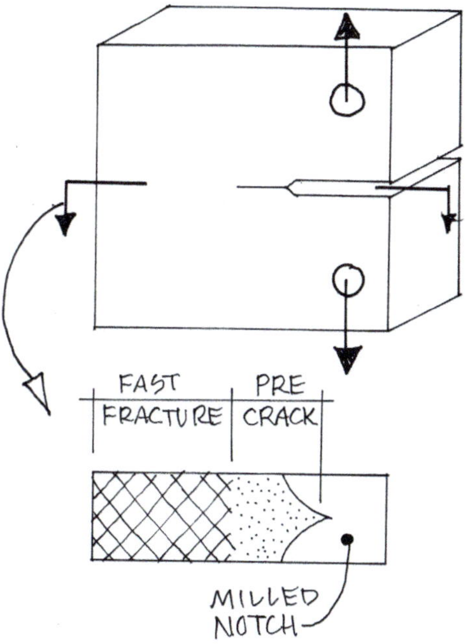

Figure 7.13 Fracture Mechanics Test Specimen

For this test to be conducted, the specimen is pre-cracked, so that it reflects the conditions of an actual crack. Then, the specimen is loaded until the crack propagates unstably. Using a **stress-intensity** solution, we can calculate the fracture toughness and apply it to design.

7.2.4 Fracture Mechanics

The most effective analytical and experimental method of predicting crack propagation is **fracture mechanics**. Design techniques could not explicitly consider cracks before A.A. Griffith proposed his theory on crack propagation in glass specimens, and Irwin made it usable and extended it to other materials in 1948. Before this time, engineers could not analytically predict at what size a crack would propagate unstably.

Fracture mechanics received its start when Griffith was trying to understand the effect of surface treatment on the strength of cyclically loaded metal parts. To reduce the potential confusion plastic deformation might cause, he began testing glass, because of its "**brittle**" behavior at room temperature. From his investigations from 1918 to 1920, Griffith proposed that a crack would propagate when the change in elastic energy with respect to crack length (dU/da) equaled the energy required for that increment of growth (dW/da). From this concept, for a linear elastic material, Griffith derived the following relationship:

$$\underbrace{f\sqrt{\pi c}}_{K} = \underbrace{\sqrt{2E\gamma}}_{K_c} \tag{7.1}$$

where:

f = far-field stress
E = elastic modulus
γ = surface tension
c = half crack length of a center-cracked specimen

When the left side equals the right side, fracture will occur. The only challenge with solving the equation is that γ is virtually impossible to obtain—so challenging, in fact, that nobody used the Griffith expression until George Irwin modified it, while at the Naval Research Lab decades later. Irwin proposed that the right-hand side of Equation (7.1) could be experimentally determined and called it fracture toughness. When the left-hand side of Equation (7.1), known as the stress-intensity factor K, equals the toughness K_c, the crack will propagate unstably (approximately one-third the speed of sound in the material). From this concept, many analysts have developed stress-intensity solutions for a wide variety of geometry and loading conditions. These are available in a multitude of handbooks.

These developments opened a new world in the prediction of fracture behavior. It was no longer based solely on past experience, and engineers could predict the behaviors of structures that hadn't been built yet. The power of fracture mechanics is that it tells the designer the size of the crack a structure can withstand before final instability. Then, using fatigue correlations, engineers can predict how long it will take for a fatigue crack to reach this size.

7.3 FATIGUE

Fatigue is time-dependent crack growth due to cyclic stresses. More rigorously, it is:

> the process of progressive localized permanent structural change occurring in a material subjected to conditions that produce fluctuating stresses and strains at some point or points and that may culminate in cracks or complete fracture after a sufficient number of fluctuations.[5]

This definition brings out seven keys to understanding that fatigue:

- is a process;
- is progressive, occurring over time;
- is localized;
- causes permanent structural change, both from cracking and load redistribution;
- is caused by fluctuating stress and strain;
- occurs at a point or points;
- results in cracks or complete fracture.

Of these points, the first one is the most fundamental to our understanding of fatigue. David W. Hoeppner, a visionary in fatigue research, stated:

> the idea that fatigue is a process is critical to dealing with it in design and to the characterization of materials as part of design. In fact, this idea is so critical that the entire conceptual view of fatigue is affected by it![6]

We separate fatigue into high- and low-cycle types. **High-cycle fatigue** is commonest, occurring over hundreds of thousands of cycles at relatively low stresses. Figure 7.14 shows such cracks in the skin of an airplane. **Low-cycle fatigue** occurs at high stresses, often above yield—as in an earthquake.

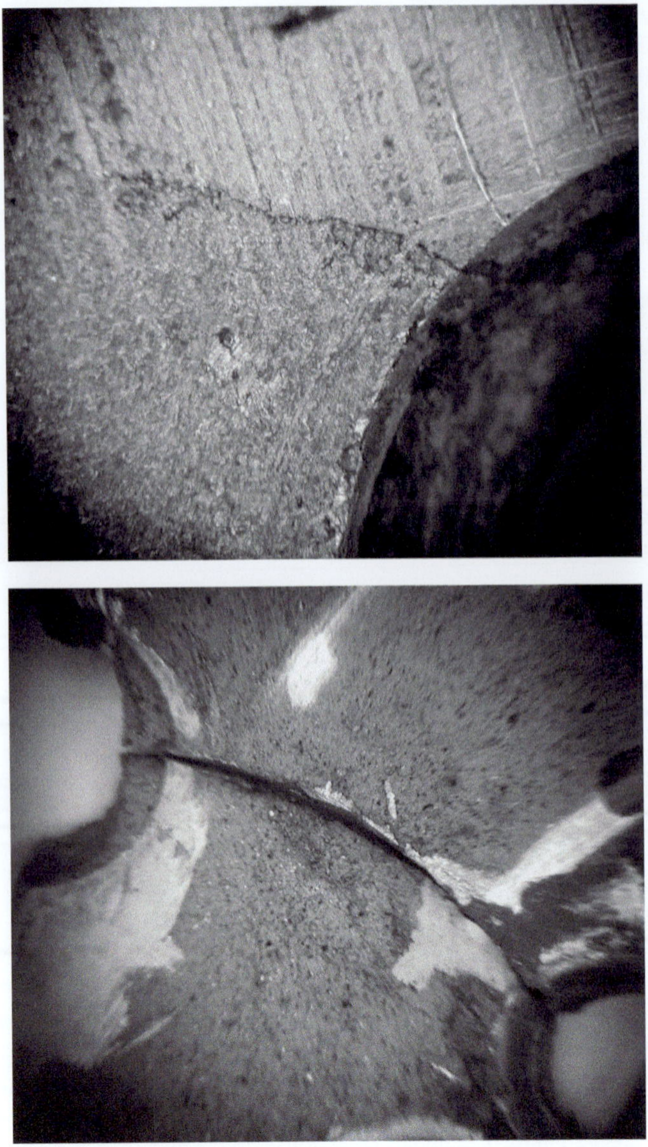

Figure 7.14 Fatigue Crack (a) Emanating from a Rivet Hole and (b) Linking Two Rivet Holes in an Aircraft Fuselage

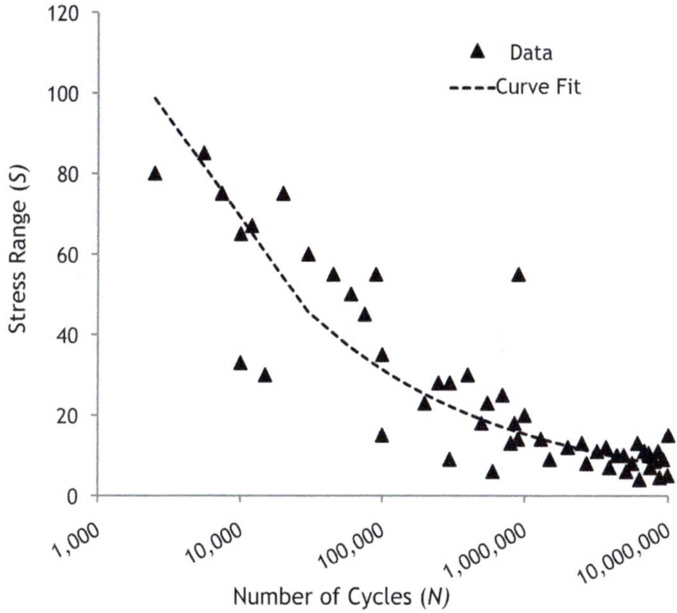

Figure 7.15 S–N Curve for Safe Life Fatigue Design

Two primary design methods of treating fatigue exist: safe life and fracture-mechanics correlations. Safe life utilizes stress–number of-cycles (S–N) curves, such as that in Figure 7.15. It operates on the faulty concept that, if one keeps stresses low enough, cracking will never occur. Not very realistic.

A more effective fatigue design approach is to use fracture-mechanics correlations. The engineer can relate the change in crack length to the stress-intensity factor range per cycle. This is accomplished through a da/dN versus ΔK curve, such as the one in Figure 7.16. The curve is based on test data and, because it is related to change in stress-intensity factor, can be extended to different component and crack geometries—serious design prowess.

By curve-fitting the data to an equation, rearranging so da and dN are on opposite sides of the equation, and integrating with respect to crack size a, we determine the total life. The distinct advantage of presenting fatigue data in this manner is that it explicitly considers initial discontinuity size.

7.3.1 Fatigue Detailing

Detailing matters. A notch in commercial construction is often not a problem, but, in a fatigue-sensitive structure, it could be catastrophic. Figure 7.17 shows simple modifications to common details that can provide substantially longer fatigue life. Notice how the changes center round the smoothing out of notches, constraint being reduced, and weld residual stresses being lowered.

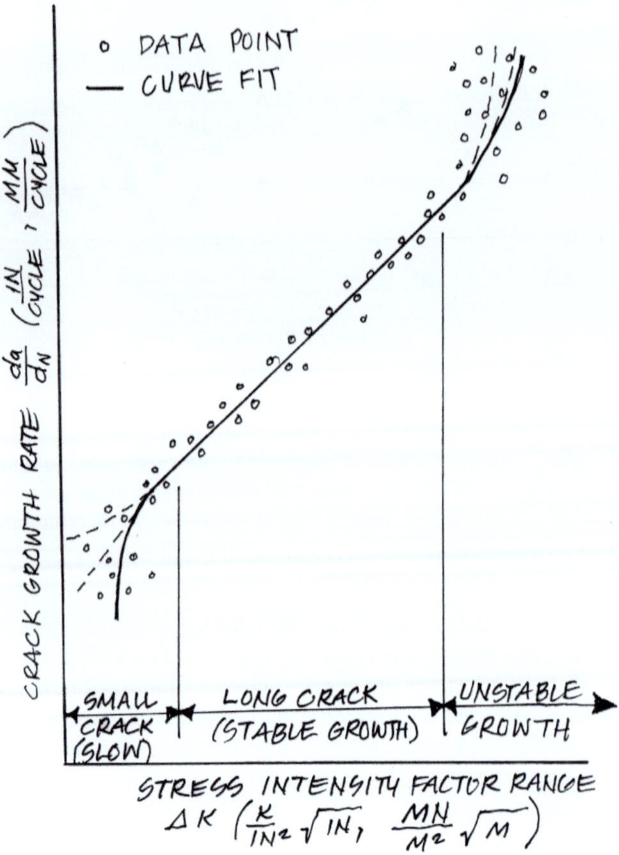

Figure 7.16 *da*/*dN* versus Δ*K* Curve for Fatigue Life Correlation

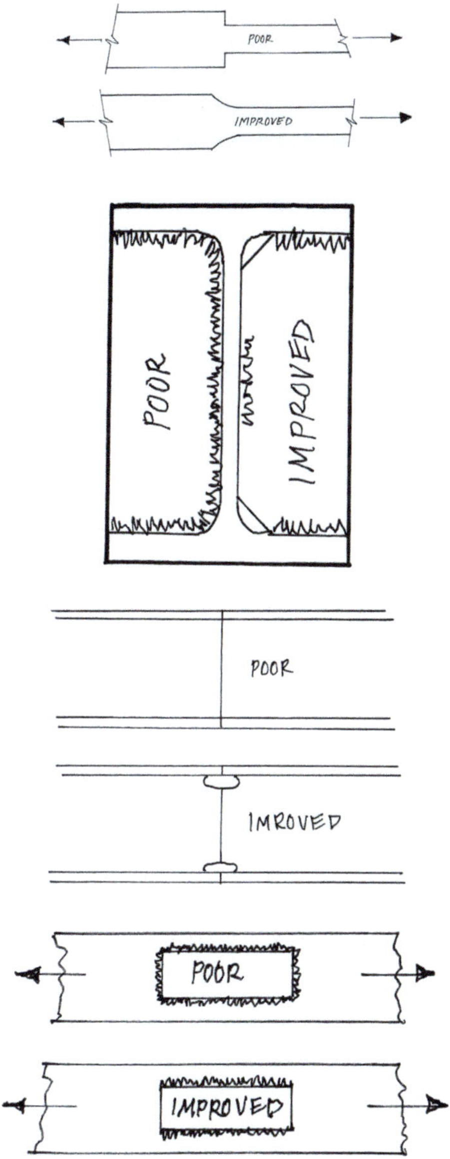

Figure 7.17 Poor and Improved Fatigue Detailing Examples

7.4 CORROSION

Because the universe favors a lower energy state, anything requiring energy to create it will desire to return to its original or even lower energy state. Corrosion is one process through which metals move to a lower energy state. The Federal Highway Administration estimates corrosion loss in the United States at 3.1 percent of the gross domestic product.[7] This equates to $555 billion a year. Chances are, moving around your city, you will see some of this loss in bridges, tunnels, and industrial facilities.

7.4.1 Corrosion Types

We often think corrosion takes place solely on large surface areas (uniform), and that its only effect is area or volume reduction: essentially, the outside gets "rusty," and the cross section gets smaller. However, this is but one of a multitude of corrosion forms and effects. Table 7.1 presents forms of corrosion found in building structures and identifies example photos.

7.4.2 Corrosion Cell Requirements

Four components are required for a corrosion cell to occur: **anode**, **cathode**, **electrolyte**, and **electron flow**. This fundamental requirement provides the basis for all corrosion prevention and protection programs. When one eliminates any of the four requirements, corrosion does not happen.

Table 7.1 Corrosion Types and Figure References

Corrosion Type	*Extents Type*	*Figure*
Atmospheric	Uniform	7.18
Exfoliation	Uniform or Local	7.19
Galvanic	Uniform	7.20
Filiform	Local	7.21
Pitting	Local	7.22
Impingement	Local	7.23
Corrosion fatigue	Local	7.24
Stress corrosion cracking	Uniform or Local	7.25

Figure 7.18 Atmospheric Corrosion of a Wide Flange Column

Figure 7.19 Exfoliation Corrosion of Bar Grating

Figure 7.20 Galvanic Corrosion in Pipe Transition between Stainless and Carbon Steels

Figure 7.21 Filiform Corrosion of a Carbon Steel Toekick

Figure 7.22 Pitting Corrosion of Steel Wide Flange Column

Figure 7.23 Impingement Corrosion of Rain Gutter

Source: Photo courtesy of NASA

Figure 7.24 Corrosion Fatigue of an Aircraft Propeller

Source: Photo courtesy of NASA

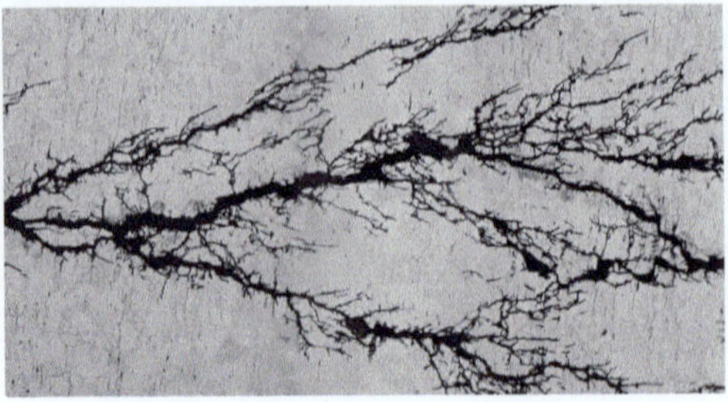

Figure 7.25 Environmentally Assisted Cracking under Sustained Tensile Load

Source: Photo courtesy of NASA

The basic corrosion cell, shown in Figure 7.26, shows the four required components for corrosion. The material corroding is at the anode, and the material not corroding (sometimes being plated) is the cathode. The electrolyte allows the metal ions to move about, and the electrical connection between the anode and cathode permits electron flow in the system.

One of the challenges in corrosion design is the multitude of ways that corrosion cells form. Although the general diagram of Figure 7.26 provides a starting point for the understanding of corrosion, it does not occur in such a clear fashion in reality (except perhaps in batteries). Figure 7.27 shows an example of the anode, cathode, electrolyte, and electron flow path in practice.

7.4.3 Energy Requirements for Corrosion

Although an anode, a cathode, an electrolyte, and an electron path are required for corrosion to take place, the simultaneous existence of these does not guarantee corrosion. The reaction must be energetically favorable. We can understand the potential for a corrosion reaction to occur by looking at the **galvanic series**, which qualitatively represents the relative potential for corrosion of metals in different solutions. An important issue that the galvanic series brings out concerns the potential changes as a function of the electrolytic environment. Table 7.2 shows the galvanic series for various metals in seawater. For further information on the kinetics of corrosion, see ASM Handbook, Vol. 13, *Corrosion*.[8]

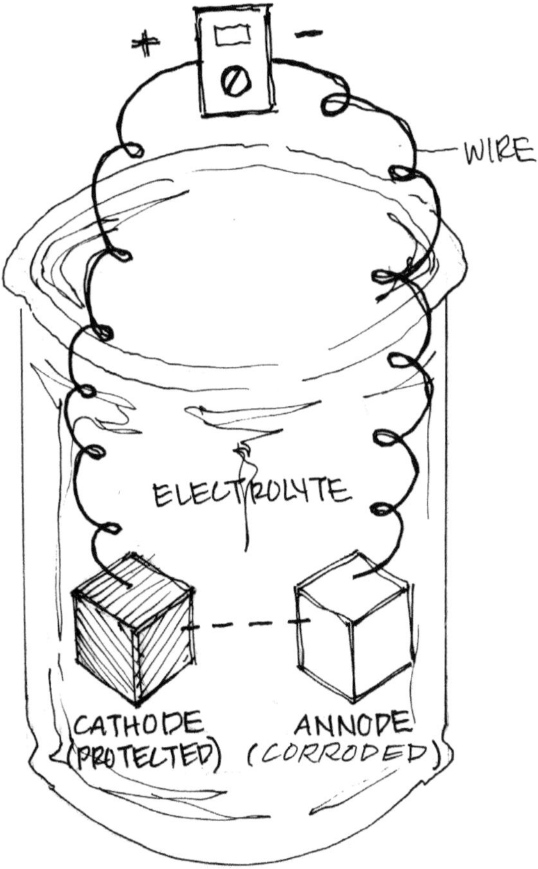

Figure 7.26 Basic Corrosion Cell

7.4.4 Corrosion-Resistant Design

Although scientists and engineers do not know everything about corrosion and its different forms, corrosion-resistant design is still possible. Three primary philosophies exist for dealing with corrosion: "ignore it," "maintain it," and "design it in." One can imagine how the "ignore it" philosophy works.

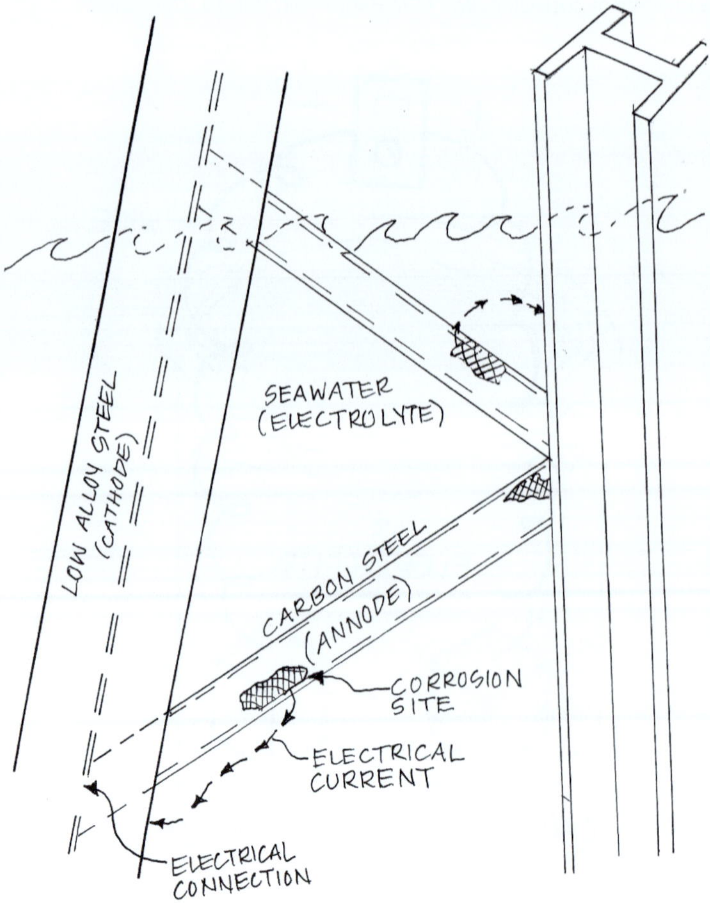

Figure 7.27 Galvanic Corrosion Cell in Marine Structure

Table 7.2 **Galvanic Series for Various Metals in Seawater**

Anodic End (Corroded)
Magnesium
Zinc
Aluminum
Low-carbon steel
Cast iron
Stainless steel
Lead
Tin
Copper
Nickel
Inconel
Silver
Titanium
Graphite
Gold
Platinum
Cathodic end (protected)

Source: Corrosion, ASM Handbook, Vol. 13

The second approach, "maintain it," is the commonest, and certainly a necessity. This essentially places the responsibility on maintenance personnel to detect and repair corrosion damage. This can be more or less effective, depending on the knowledge and resources available for maintenance.

The third paradigm is to "design it in." This paradigm rightly assumes that designers can avert the majority of corrosion problems by proper design. This design includes the following parameters:

- material selection;
- material protection;
- construction details;
- construction procedures;
- adequate inspection and maintenance specifications.

Though it is beyond the scope of this book, one can begin to see how each item in the list provides the framework for a robust design and maintenance approach.

7.4.5 Corrosion Detailing

Let's look at a few simple detailing modifications that reduce corrosion. The focus is on reducing or eliminating locations where debris and

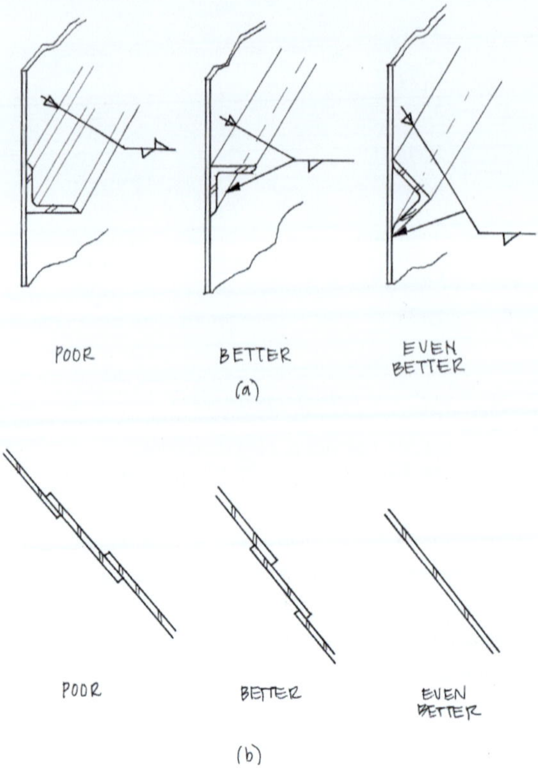

Figure 7.28 Corrosion-Resistant Details

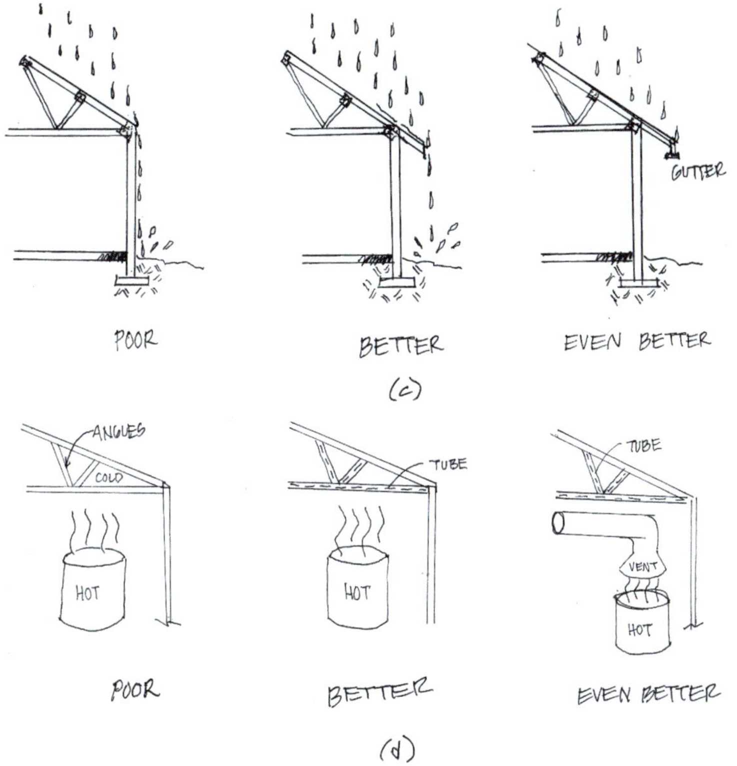

Figure 7.28 Continued

moisture can collect, or on configuring the structure so that it does not splash on itself, as illustrated in Figure 7.28. Consider how you can expand these concepts to other conditions.

7.5 CONCRETE DETERIORATION

Well-designed and -constructed concrete structures can last decades, even centuries, with little maintenance. However, a number of mechanisms can hasten their demise. There follow three of the commonest, together with a discussion on how to reduce the likelihood of their occurring.

Alkali–silica reactivity occurs between the highly alkaline cement paste and non-crystalline silica. It leads to alligator-like cracking and strength reduction (see Figure 7.29). It has led to the demolition of numerous structures. To protect against this, batch plants test aggregate and cement mixtures to ensure this reaction will not occur. Where it does, fly ash or silica fume can reduce the reaction potential in new mixes. In existing structures, replacement is often the only option.

Sulphate attack occurs when sulphates in soil or seawater degrade the cement paste. This reaction forms expansive by-products and leads to cracking and scaling. Sulphate attack manifests itself in cohesion and strength loss in the concrete (Figure 7.30). Type V cement is commonly used and very effective in resisting sulphate attack. Silica fume densifies the concrete and also helps.

Chloride penetration occurs when concrete is exposed to seawater or de-icing salts and is not sufficiently dense. Chlorides migrate to the reinforcing steel and break down the passive barrier created by the cement. This causes corrosion, leading to volume change in the steel and resulting in cracking, loss of cover, and reinforcing steel loss (Figure 7.31).

Figure 7.29 Alkali–Silica Reaction in Concrete

Figure 7.30 Sulphate Attack in Concrete Wall

Figure 7.31 Chloride-Induced Corrosion in a Retaining Structure

Concrete mix design, reinforcing cover, and coatings are options to resist reinforcing steel corrosion.

Mix designs that densify concrete limit chloride penetration. Silica fume and fly ash are effective in densification. Ensuring the mix water and admixtures don't contain chlorides is also important. Also, providing sufficient cover between the bar and concrete surface is critical to protecting the reinforcing steel. Cover thickness ranges from ¾ in to 3 in (19–75 mm). Finally, galvanizing or epoxy-coating reinforcing steel can be a good last resort. However, if the first two remedies are well executed in the field, bar coatings are usually unnecessary.

7.6 CREEP

Creep is time-dependent, permanent deformation due to sustained loads. It occurs in timber, concrete, masonry, and plastics at room temperature, and in metals at elevated temperature. It causes additional deflection over time. Take a plastic bottle, fill it with water, and hang it somewhere, and you will see creep firsthand.

7.7 WEAR

Wear is damage to a solid surface due to relative motion between the surface and a contacting material. It results in progressive material loss. Wear doesn't occur much in traditional commercial buildings, but is prevalent in manufacturing and industrial facilities. Options for wear resistance include:

- lubrication;
- sacrificial materials;
- hardened materials at wear locations.

7.8 HOLISTIC STRUCTURAL INTEGRITY

Our tendency as humans is to simplify problems so they are manageable. Much of engineering goes in this direction. Although it helps us to move forward on a project, we sometimes miss things that matter. The treatment of structural integrity is immensely complex and equally filled with oversimplifications. If we recall some of the inputs in Figure 7.2, it becomes immediately apparent that we need a broad framework. Enter **holistic** structural integrity.

D.W. Hoeppner and R.H. Jeal, in the 1960s, began to see a need for a broader view of fatigue and fracture. Building on the work of A.A. Griffith,

P.J.E. Forsyth, W.A. Wood, and A.M. Freudenthal, they proposed the four stages of crack growth we previously saw in Figure 7.6. They then extended it to consider corrosion, creep, and wear, and their synergistic effects. This was novel thinking then, and still is today. When we approach structural integrity holistically, we have an experiential, experimental, and analytical framework for managing all the variables. We then ask the right questions and put the answers in reasonable relationship to each other.

Key components of the framework are:

- Characterizing material behavior in the service environment: This includes corrosion behavior, fatigue crack growth rates, or fracture toughness.
- Understanding the initial and in-service discontinuity state: This is done through quality control and inspection. These discontinuities may be cracks, corrosion damage, wear, or creep. Based on the threshold of detection, we can assume these discontinuities exist in the most critical location of a member and design for it.
- Predicting the life of the structure: We do this with structural analysis, fatigue correlations, corrosion and wear estimates, and fracture mechanics.

Putting these concepts into practice for fatigue life prediction, we see the following emerge:

- Develop a load spectrum, capturing steady-state and overload cycles.
- Characterize the fracture toughness and crack growth rate for the material, environment, and loading.
- Understand the fabrication and inspection tolerances of what we are designing.
- Assume there is a crack at the most critical location, the size of our inspection tolerance.
- Predict the critical crack size using fracture mechanics.
- Integrate the crack growth curve (Figure 7.16) from the inspection threshold to critical crack size to predict life.
- Develop an inspection plan, as discussed in the next section, to close the information loop.

7.9 INSPECTION

Inspection closes the loop for holistic structural integrity. It allows the engineer to know what a structure's initial discontinuity state is owing to fabrication, and to evaluate changes as the structures ages. It is, therefore, critical that we have a rational and robust inspection plan.

Table 7.3 Nondestructive Inspection Thresholds of Detection

Test Method	*Discontinuity Sizes*			
	Minimum		*Maximum*	
	(in)	*(mm)*	*(in)*	*(mm)*
Surface				
Liquid penetrant	0.017	0.43	0.700	17.78
Magnetic particle	0.039	0.99	0.237	6.02
Eddy current	0.022	0.56	0.750	19.05
Internal				
Ultrasonic	0.014	0.36	0.265	6.73
Radiographic	0.024	0.61	0.729	18.52

The key components of any inspection plan are:

1. what to look for;
2. when to look;
3. how to look;
4. where to look;
5. how often to look;
6. threshold of detection;
7. probability of detection.

Let's review briefly how to look. Nondestructive test methods can be broken into two groups: surface and internal, as shown in Table 7.3. Each group has a unique place and ability to find discontinuities. Magnetic particle and ultrasonic testing are the commonest methods used in civil structures, to detect surface and internal cracks, respectively. Coupling inspection technique with threshold of detection, we can know what our initial crack size is for design.

7.10 CONCLUSION

Structural integrity is often where traditional structural engineering breaks down. It is complex. Because of this, we tend to oversimplify it, which leads to inadequate models and understanding of what is really going on. However, when we open our minds to what may happen to our structure, we are prepared to handle the challenges.

Traditional design methods oversimplify and disjoint the effects of fracture, fatigue, corrosion, concrete deterioration, creep, and wear mechanisms. Holistic structural integrity brings them all back together where they belong, coupling them with testing, analysis, and inspection tools. This provides us a robust, rational framework to design for the universe's tendency to a lower energy state—resulting in safer structures.

NOTES

1. H. Petroski. *To Engineer Is Human* (New York: First Vintage Books, 1992).
2. AISC. *Specification for Structural Steel Buildings*, AISC 360 (Chicago, IL: American Institute of Steel Construction, 2010).
3. P.W. McMullin. "General Principles of Fatigue and Fracture, Part 1," *Structure Magazine*, August (Reedsburg, WI: Copper Creek, 2016).
4. W.S. Pellini. *Evolution of Engineering Principles for Fracture-Safe Design of Steel Structures*, NRL Report 6957 (Washington, D.C.: Naval Research Laboratory, 1969), 87.
5. ASTM. *Definitions of Terms Relating to Fatigue*, ASTM E1150 (West Conshohocken, PA: ASTM International, 1993).
6. D.W. Hoeppner. "Industrial Significance of Fatigue Problems," in *Fatigue and Fracture*, ASM Handbook, Vol. 19 (Materials Park, OH: ASM International, 1996), 3.
7. FHWA. *Corrosion Costs and Preventative Strategies in the United States*, Report FHWA-RD-01–156 (Washington, D.C.: Federal Highway Administration, 2002).
8. D.W. Shoesmith. "Kinetics of Aqueous Corrosion," in *Corrosion*, ASM Handbook, Vol. 13 (Materials Park, OH: ASM International, 1987), 29–36.
9. W.S. Pellini. *Principles of Structural Integrity* (Arlington, VA: Office of Naval Research, 1976).

Cladding

Chapter 8

Justin W. Jacobs and Paul W. McMullin

8.1 INTRODUCTION

Cladding provides the physical barrier between us and the elements. Without it, building occupancy during times of environmental extremes would be unpleasant at best. Architectural cladding consists of almost any imaginable combination of transparent, translucent, and opaque components. Transparent glazing lets light in while permitting views out and provides a necessary connection between building occupants and the outside world; translucent glazing still lets light in, but provides some amount of privacy. Opaque elements provide architectural interest, better insulation, and enhancements to the façade's durability.

As sustainable buildings gain greater societal importance, cladding for both energy efficiency and openness has required additional thoughtfulness in its design. It is not necessary for a building to be ugly to be energy efficient. It just takes greater resources and social impetus to have both.

8.1.1 Who Is Responsible?

Cladding design responsibility lies foremost with the architect, but often requires an interdisciplinary team of architect, mechanical and structural engineers, lighting designer, specialty cladding designer, specialty structural engineer, and installation subcontractors. For simple cladding types, the architect and engineer often work out the details. For high-performance structures, a focused specialty team can bring great energy to the project, along with visual, financial, and performance benefits.

The interface between cladding and building structure is often a grey area of design responsibility. Most structural engineers focus their efforts on the building **superstructure**, whereas architects are concerned primarily with the aspects of aesthetics and weather intrusion. Addressing cladding early in the project yields financial and schedule benefits because it is not an afterthought. Clear contracts and early, thorough coordination can alleviate many challenges associated with this reality.

8.2 CLADDING TYPES

8.2.1 Glazing

Glazing makes up a significant portion of the **envelope** of common buildings and interfaces with all other cladding materials. Commercial buildings today generally utilize dual-pane insulated glazing systems for energy efficiency. However, triple-pane systems are becoming more common in highly efficient buildings.

A typical dual-pane insulated glazing unit (IGU) is made up of two ¼ in (6 mm) thick layers of glass separated by a ½ in (12 mm) air space, connected together around the perimeter of the unit with a **nonconductive** spacer, illustrated in Figure 8.1. The performance of an IGU is measured in several ways, thermal transfer, visible light transmittance, and solar heat gain being the most important. Thermal performance can be improved by displacement of the air between the panes of glass with a noble gas such as argon or xenon, or by the addition of a third layer of glazing to the system. To reduce visible light transmittance, providing for a better level

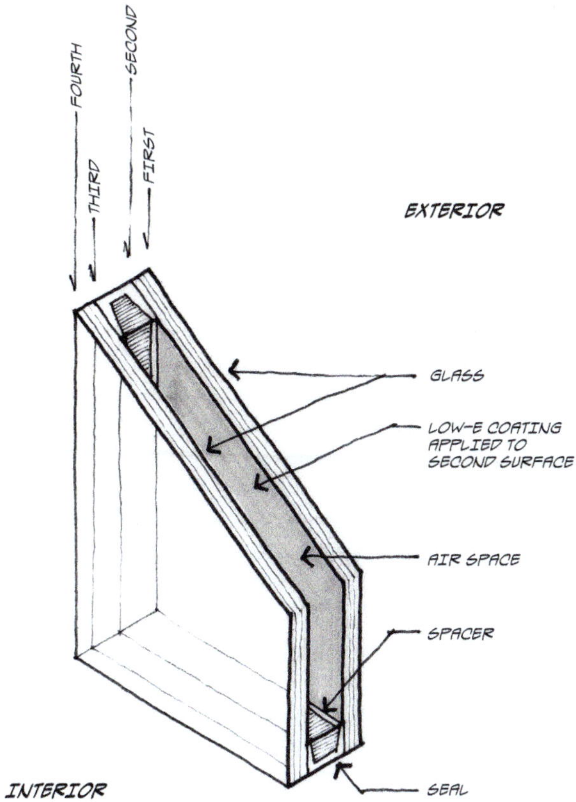

Figure 8.1 Cutaway Section of Multi-pane Glazing with Spacer

of occupant comfort by reducing glare, one or both panes of glass can be tinted with the inclusion of an applied film. By far the most popular method for increasing the performance of glazing is the installation of a **low-e coating**, which limits the infrared component of sunlight passing through the window, on the second or third surface of the glazing.

In commercial construction, exterior openings in opaque walls are typically installed as storefront framing systems or curtain walls, both of which are made up of aluminum extrusions connected together in grids. Storefront systems are limited to single-story applications and can be used as either ribbon or punched openings. Aluminum curtain walls are capable of spanning multiple floor levels, as they are made up of structural vertical members and are constructed with provisions to accommodate movement and moisture within the system, as shown in Figure 8.2.

As with insulated glazing, there are ways to specify frame products or materials that improve the performance of the overall system. The simplest, least efficient of these **extrusions** is a non-thermally broken frame piece (Figure 8.3) that allows thermal energy to move freely between the interior and exterior of the building. If a nonconductive thermal spacer is inserted between two separate extrusions, the interior and exterior environments can be separated, which reduces thermal transfer and **condensation** between the faces of the frame (Figure 8.4).

Of the many types of glazing available, building envelopes today overwhelmingly use one or more of these three: heat-strengthened (HS), tempered, and laminated. Each of these three types of glazing begins life as annealed glass, which is fabricated using a float process consisting of heating silica sand and other minerals to 2,700°F (1,500°C) and "floating" the resulting fluid over a layer of molten tin. This fluid glass is then cooled on a conveyor and cut into large panes to be processed again.

HS glass is annealed glass that has been cut to size, heated to approximately 1,200°F (650°C), and then cooled at a controlled rate. The resultant pane of glass is approximately twice as strong as annealed glass and can hold up to thermal shock and stress induced by the environment (e.g., uneven shading, back painting, or back insulating) much better in service than annealed glass. Fully tempered (FT) glass is made in a very similar way, but the resulting pane is approximately twice as strong as HS glass and will break into very small pieces when broken. For this reason, FT glass is required in passage doors, tub and shower enclosures, and where installed adjacent to walking surfaces.

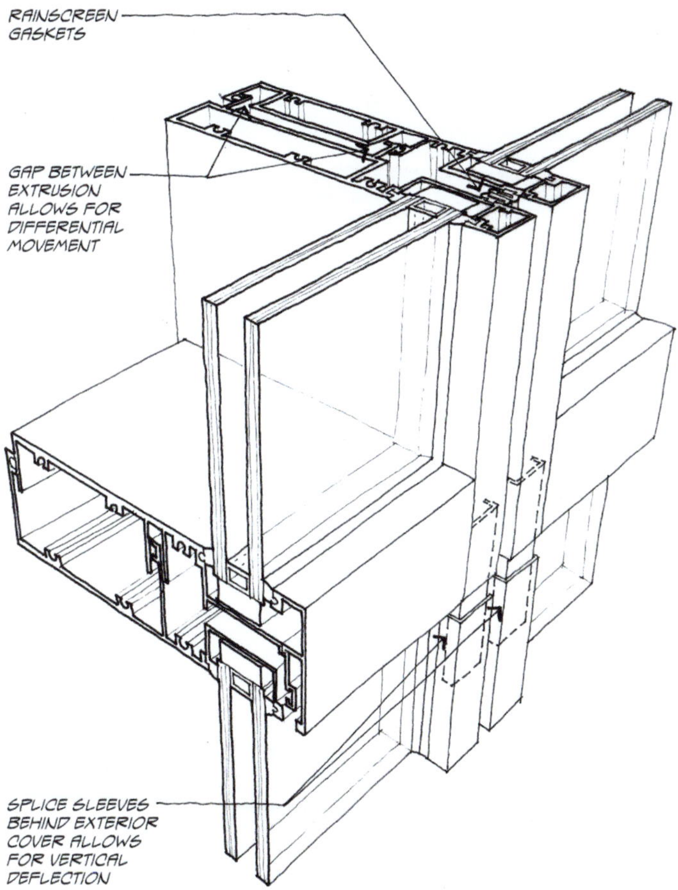

Figure 8.2 Vertical Curtain Wall Movement Joint

Laminated glass consists of two or more layers of glass that have been chemically bonded together with a transparent plastic interlayer, illustrated in Figure 8.5. It is used in places where added strength or even bullet and blast resistance is required and is commonly found in automobile windscreens.

ASTM C1048[1] provides guidance on the proper specification and use of HS and FT flat glass. ASTM C1172[2] provides guidance on the proper

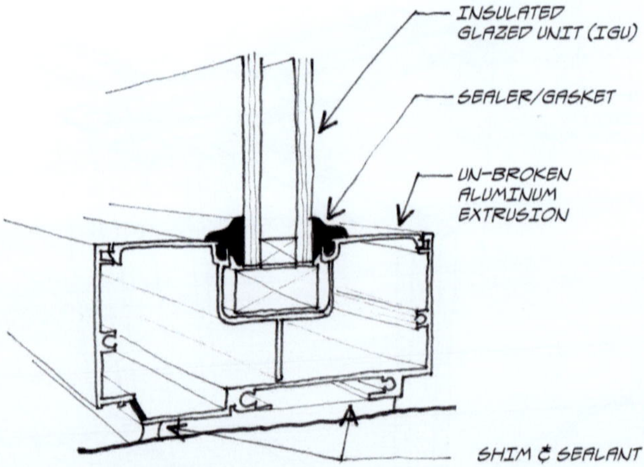

Figure 8.3 Aluminum Mullion with No Thermal Spacer

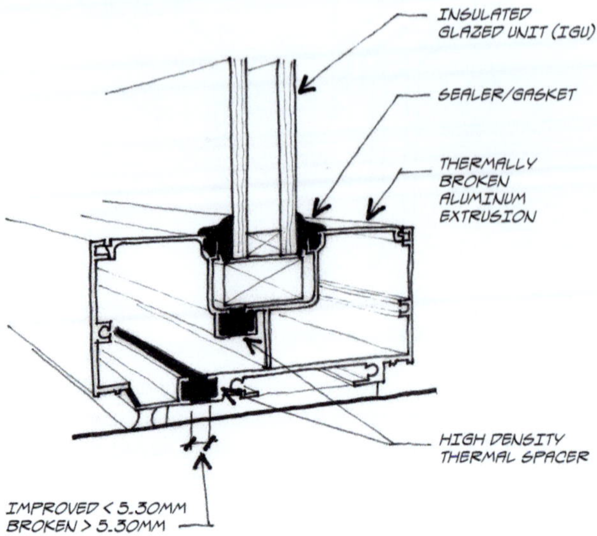

Figure 8.4 Aluminum Mullion with Thermal Spacer

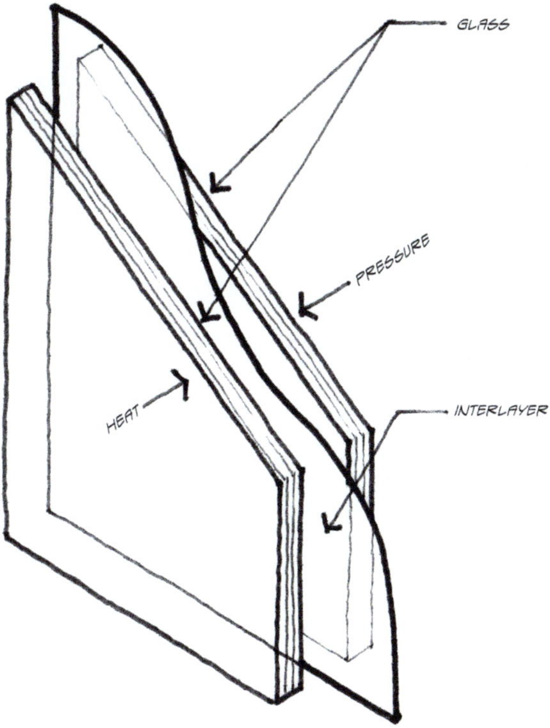

Figure 8.5 Laminated Glass with Interlayer

specification and use of laminated architectural flat glass. The figures provided in ASTM E1300[3] provide glazing thickness as a function of span and pressure.

8.2.2 Metal Studs

Opaque wall elements are often backed by metal stud walls and sheathing in modern commercial construction. These can be designed to span vertically, either from one floor to the next (platform method), as shown in Figure 8.6, or across multiple levels (balloon method), illustrated in Figure 8.7. It is imperative that consideration be given to the size, weight, expansion/contraction potential, construction **tolerances**, story **drift**, and installation methods of the intended cladding when the appropriate framing method is determined, as well as the size, gauge, and spacing of

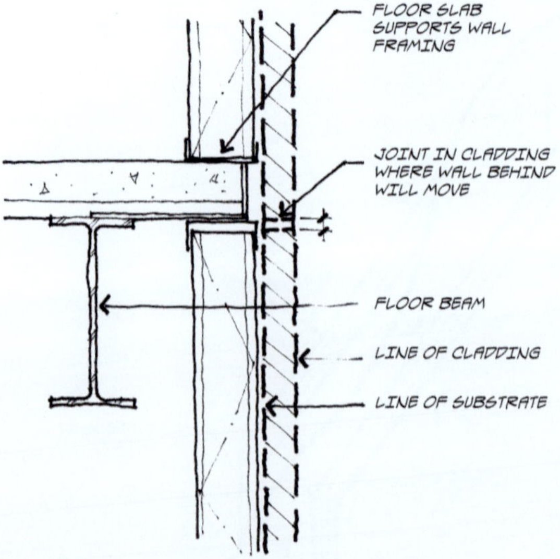

Figure 8.6 Platform Framing Methods at Typical Floor Level

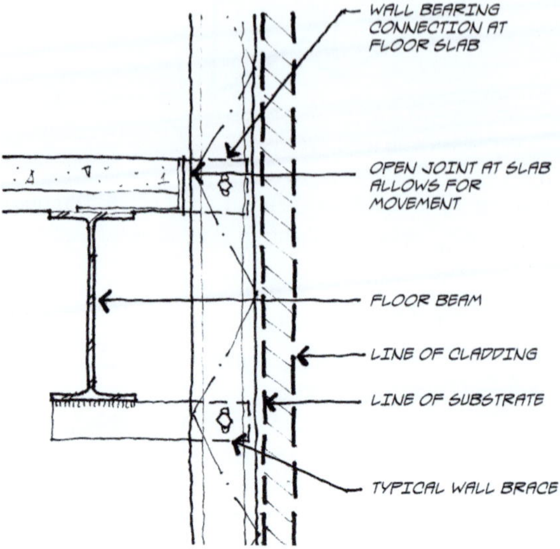

Figure 8.7 Balloon Framing Methods at Typical Slip Floor Level

the metal studs themselves. Other factors to consider are building height, opening sizes and locations, and prevalent local construction methods.

Balloon framing is capable of accommodating floor slab **deflection** (vertical), as well as seismic or wind movement (horizontal), by attaching directly to the floor slab, as illustrated in Figure 8.7. This framing method allows the cladding to "ride along" with the floor slab when it deflects. When platform framing methods are used, cladding details must accommodate this movement with horizontal joints in the system.

When choosing a framing system, consider the challenges that will arise during construction. For instance, how will the framer access the top track of a platform framed wall, where there is likely a spandrel beam in the way? Conversely, how much additional material and labor might be required in a balloon framing condition in order to brace the wall back to the structure and prevent **racking**?

There are many published standards regarding the specification and design of metal studs. Primarily, ASTM C645[4] is the industry standard for the manufacturing of nonstructural metal framing, and ASTM C754[5] provides guidance for installation of nonstructural metal framing, including attachment methods, span tables, and limiting heights.

8.2.3 Curtain wall

Curtain wall is a nonstructural building skin consisting of aluminum mullions in a horizontal and vertical grid, with an infill of glazing or light-weight opaque panels such as: stone veneer, metal panels, louvers, and operable windows or vents. Vertical mullions in these systems tend to be continuous over two floors, laterally supported at each floor, and supported vertically at every other floor. This creates a need to detail the vertically unsupported connections to slip (see Figure 8.2).

Pay close attention where a curtain wall spans multiple levels within a building. In this case, we must design the curtain wall to accommodate story drift, which is primarily addressed through racking of the frames, as shown in Figure 8.8. As the shape of glass or opaque panels will remain unchanged in this condition, there must be enough space between their edges and the interior edge of adjacent mullions to accommodate this movement. It is especially critical at the corners of the building, where horizontal drift and tilting occur together, as illustrated in Figure 8.9.

Aluminum curtain wall can be designed and installed using one of two basic methods: stick (ladder) and unitized. Stick construction consists of

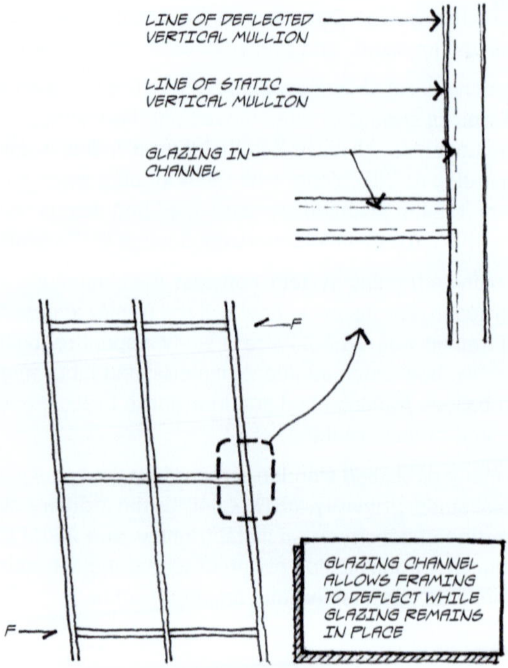

Figure 8.8 Curtain Wall Drift Accommodation

aluminum mullions (sticks) installed vertically and then horizontally between the vertical ones. The glazing then fits in the openings within the grid, illustrated in Figure 8.10. In this method, the vast majority of the work is performed at the jobsite.

Unitized construction offers a much more efficient final product, but requires far greater design effort. As its name implies, unitized curtain wall is made up of somewhat regularly sized/shaped factory-built panels, illustrated in Figure 8.11, that are subsequently shipped to the jobsite and hung from the building superstructure. As the majority of the work is performed in a factory, it can be performed quickly and with higher quality.

There are a great many standards we should be aware of when specifying and detailing aluminum curtain wall and storefront. Some of these standards include: ASTM E330[6] (wind load), ASTM E283[7] (air infiltration), ASTM E331[8] and **AAMA** 501.1[9] (water penetration), AAMA

501.4[10] (story drift), AAMA 501.6[11] (seismic performance), and AAMA 501.5[12] (thermal cycling).

8.2.4 Masonry

Although masonry includes virtually any **modular** building material, stacked or laid and bound together with **mortar**, this text will concentrate on brick veneer. Brick veneer falls into one of two categories, anchored or adhered, each offering their own advantages.

Adhered masonry veneer consists of a thin layer of stone or clay masonry, set directly against the exterior wall with a layer of mortar over

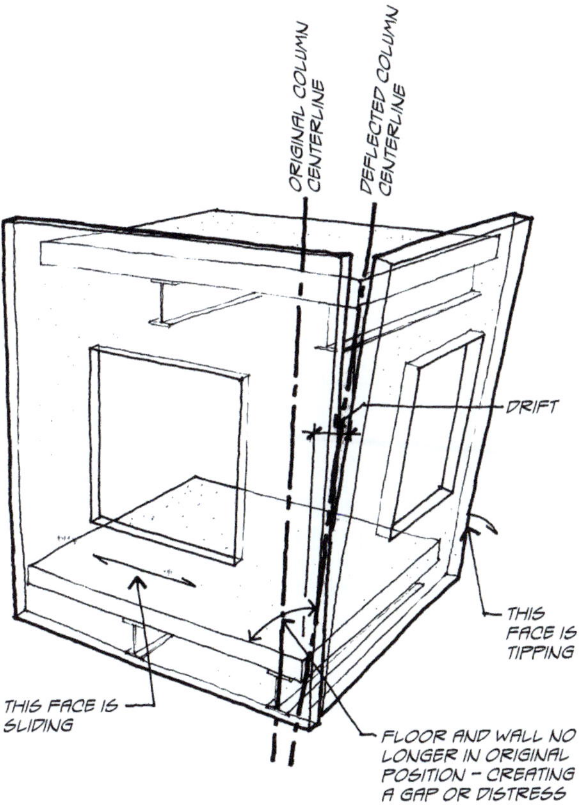

Figure 8.9 Effect of Story Drift at a Building Corner

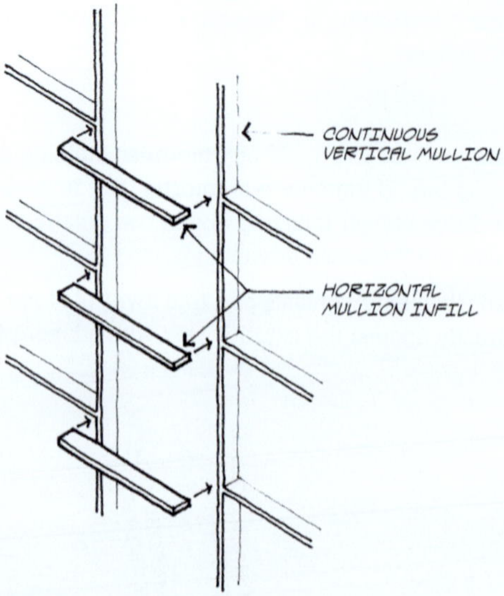

Figure 8.10 Exploded View of Stick Construction

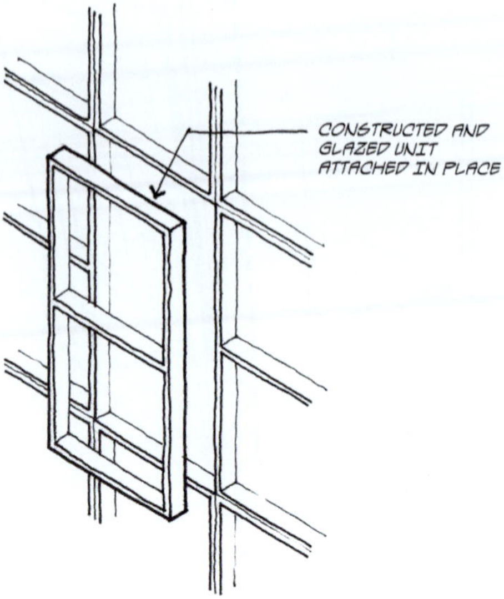

Figure 8.11 Related View of Unitized System

expanded metal lathe. This allows for simple detailing of air and moisture barriers, as well as movement. Expansion and contraction of the veneer have to be accommodated by the veneer being broken into panels that correspond directly with joints in the wall behind, as shown in Figure 8.12. These joints should be located at wall corners, changes in height, changes in wall material, and at no more than 15 ft (4.5 m) from center in either direction.

Anchored masonry, as is suggested in its description, is made up of individual clay, stone, or concrete units, stacked vertically and supported horizontally at specific intervals, as illustrated in Figure 8.13. Given the requirement for direct anchorage to the backing structure, it is imperative that the structural engineer provide the architect with guidelines for the

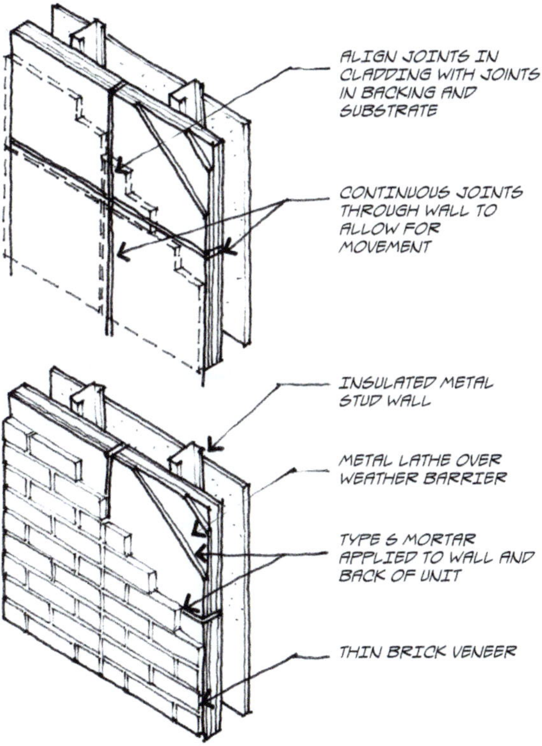

Figure 8.12 3D View of Adhered Masonry Veneer

placement of this anchorage as soon as possible in design. Although a very simple cladding option for single-story construction, anchored masonry veneer can create significant challenges for the design team to address when used in multistory construction. Expansion and contraction joints must be detailed to allow for vertical and horizontal movement in concert with the backing structure, and thermal movement and moisture growth also need to be accommodated. Clay brick will absorb water, even atmospheric moisture in dry climates, and will expand slightly as it does so.

A successful design for anchored masonry veneer will rely heavily on early, in-depth coordination between the architect and structural engineer, to ensure all required anchoring, anticipated movement, and continuity of the weather barrier have been thoroughly integrated and detailed.

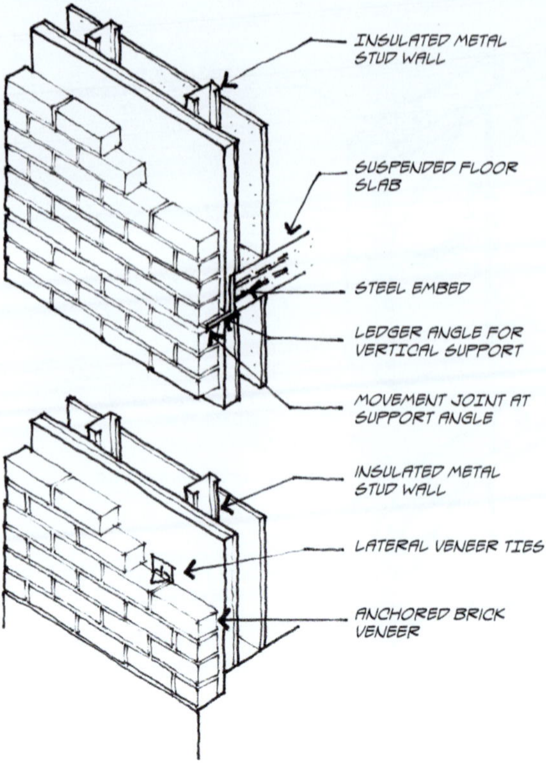

Figure 8.13 3D View of Anchored Masonry Veneer

ASTM Standards C1405[13] and C1088[14] can be helpful for the specification of brick veneers for use in different conditions.

8.2.5 Precast Concrete

Precast concrete cladding consists of large, **cementitious** panels that are formed and cast off-site in a controlled environment, then shipped to the jobsite for installation. Common in large office and municipal buildings, as well as some industrial applications, precast cladding is very durable and can be installed quickly. As one of the major benefits of precast panels is regularity, they are generally designed to span vertically, with openings cast in the panels to accommodate doors and windows (*punched*), or horizontally, with windows and doors installed between panels (*ribbon*).

Panel size is generally limited in one direction to 12 ft (3.66 m) to facilitate over-the-road transport. Consideration should be given in design to the overall size and weight of each panel. The larger the panel, the higher the transportation and erection costs, whereas smaller panels require more manufacturing and erection time. It's a delicate balance that should be discussed with the builder during design to best serve the needs of the owner.

Precast cladding can act on its own as a primary air and weather barrier, where only the horizontal and vertical joints between panels need be sealed against the environment. This provides a saving to the owner in material and labor costs. However, the panels themselves can be quite heavy and will almost assuredly require a crane to lift them into place, as well as specialized installers who are accustomed to working with these materials and, often, blind attachment methods, as shown in Figure 8.14.

Precast concrete can save significant time and money when used in relatively small areas of a building façade. For instance, as a secondary accent material on a primarily masonry-clad building, precast elements will often be used as headers over openings, column and beam wraps, **fascia**, and cornices. Detailing of these elements will be similar to the adjacent masonry in terms of allowing for building movement and thermal expansion/contraction, as discussed in Section 8.3.3.

8.2.6 Ceramic, Stone, and Cast Stone

In institutional and commercial buildings, ceramic and stone are common for their durability and elegance. Historically, these materials were the fundamental building blocks or placed over solid brick walls. Today's lighter structures, coupled with high-cost materials, require backup frames

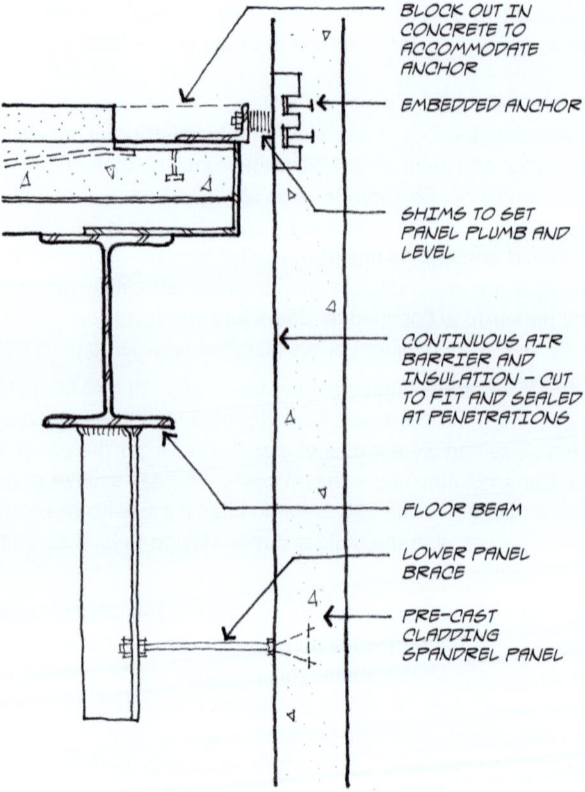

Figure 8.14 Section Detail Showing Typical Precast Concrete Anchoring

for ceramic and stone veneers, such as that in Figure 8.15, which demand coordination to ensure that the building superstructure, wall system, and attached cladding are able to react to changing conditions without failure.

Ceramic tile cladding can be adhered or anchored to the structural backing like masonry veneer. When adhered, ceramic tile cladding requires **sealant** and backup at all open joints between tiles. Although this method has a lower installed cost than that of an open-jointed anchored tile, these joints will require regular inspection and maintenance for the life of the building, increasing the overall life cycle cost of the system.

Cast stone has been a major building material for almost a thousand years and is closely related to precast concrete, with the primary difference being the type of **aggregate** used in its manufacture. If a fine aggregate is used in the mix, the final cast stone product much more closely resembles a natural rock or stone than it does concrete. The industry standard for physical properties and raw materials in architectural cast stone is ASTM C1364.[15]

8.2.7 Emerging Cladding

In addition to the tried and true building materials described above, there is always a push to bring new materials to market in an effort to provide performance gains and cost benefits to building owners worldwide. Terracotta, ipe, and porcelain tile are a few examples of currently emerging trends in cladding materials.

Terracotta is a kiln-fired clay product that originally gained popularity as an accent to brick in cladding in the late nineteenth century. In the past few decades, terracotta has shrugged off its role as an accent to become a

Figure 8.15 Galvanized Steel Backup-Type Frame for Stone Cladding

Source: Photo courtesy Kepco+

popular primary building cladding material in commercial, medical, and municipal buildings around the world. Light-weight, natural, durable, pre-finished, and long lived, terracotta checks several boxes in the list of benefits one looks for in building cladding. Another benefit to terracotta is that it lends itself to installation as a rainscreen.

Rainscreen cladding is becoming more popular, as new installation methods and materials are being developed. A rainscreen essentially opens up the exterior surface material to allow for pressure equalization between the backup air space and the external environment. Cladding is installed with open joints (see Figure 8.16) that are designed to shed the vast majority of rainwater, but allow air movement behind the cladding to promote ventilation. In contrast to traditional installation methods, where joints in the cladding between different tiles, panels, or bricks need to be filled to prevent the passage of moisture, there is no need for sealants in a rainscreen installation.

Ipe is a species of hardwood that can be found in Central and South America. Sometimes referred to as Brazilian walnut, ipe is one of the densest hardwoods on the market (300 percent harder than cedar), making it much more able to hold up to the elements than other hardwood varieties. Ipe is generally installed in plank form, either vertically or horizontally, and either lapped or locked together (Figure 8.17) directly

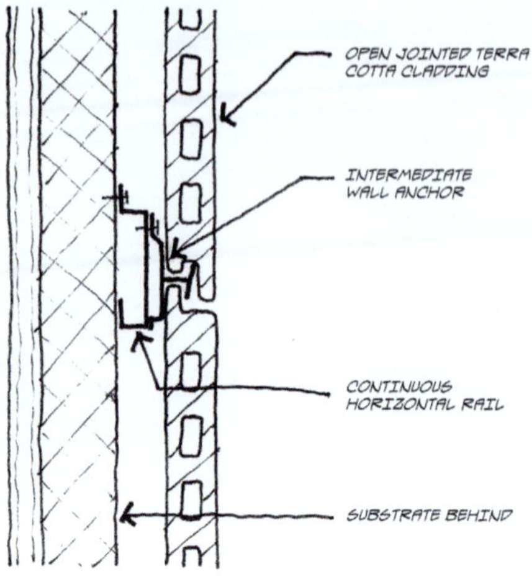

Figure 8.16 Section Detail of Baffled Terracotta Joint

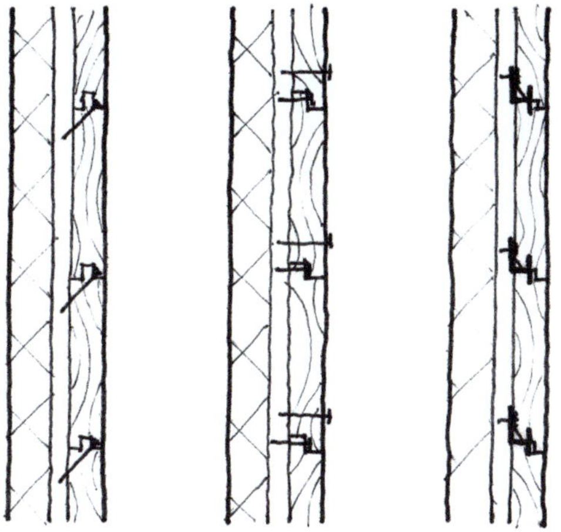

Figure 8.17 Lapped and Locked Joints

over a weather barrier on the substrate. It is naturally resistant to insect attack, maintenance-free, strong, and stable and will weather to a natural and even silvery grey over time.

8.3 ROOM TO MOVE

Because cladding is subject to the full range of environmental conditions, and the building moves during wind and seismic events, it is critical that all cladding be designed to move relative to itself as well as the primary building structure. Vertical floor and roof movements can induce loads on the cladding, and horizontal drift will push and pull the exterior envelope out of plane. These normal building movements must be accommodated, not only in the building structure, but in the cladding system as well.

In Figure 8.18, we see a building with poor and good detailing. The poorly detailed building in Figure 8.18A doesn't have room for things to move. In area 1, there are no relief joints at each floor, and yet it is rigidly attached. As the brick swells, it has nowhere to go. In area 2, there is cracking between windows because there is no slip plane allowing horizontal movement between stories. Finally, at the corners, area 3, as the wall racks, the perpendicular wall tips when the floors move relative to each other. When this corner is rigidly connected, the movement causes cracking. At best, these conditions lead to an irritated owner; at worst, someone could get hurt from falling cladding.

Compare this with the well-detailed structure in Figure 8.18B. In area 1, we have included a relief angle and flexible joint at each story, allowing thermal and moisture movement. For area 2, there are no longer cracks between windows, because we have detailed them to slide relative to the brick above. In area 3, we have a pronounced column panel that allows the brick to slide behind it. The changes are subtle; the performance is drastically different. Let's get into this further.

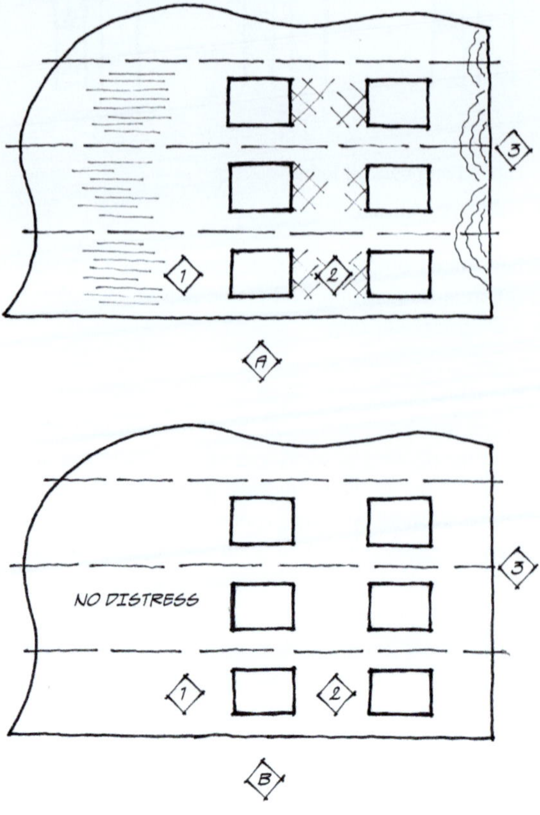

Figure 8.18 Poor and Improved Detailing for Cladding Movement

8.3.1 Expansion and Contraction Related to Thermal Change

Most commercial structures maintain a small interior thermal range compared with the outside environment. Coupled with the different thermal expansion rates of materials, this creates a condition where the cladding expands and contracts to a much greater degree than the primary building structure. The forces generated when this happens are unmanageably high, and so we design the interface to slip. Where possible, the most efficient method to provide for this movement is horizontally, at the heads of storefront ribbon windows, and vertically, at locations where materials change or structure exists, as shown in Figure 8.19. Alternatively, where a

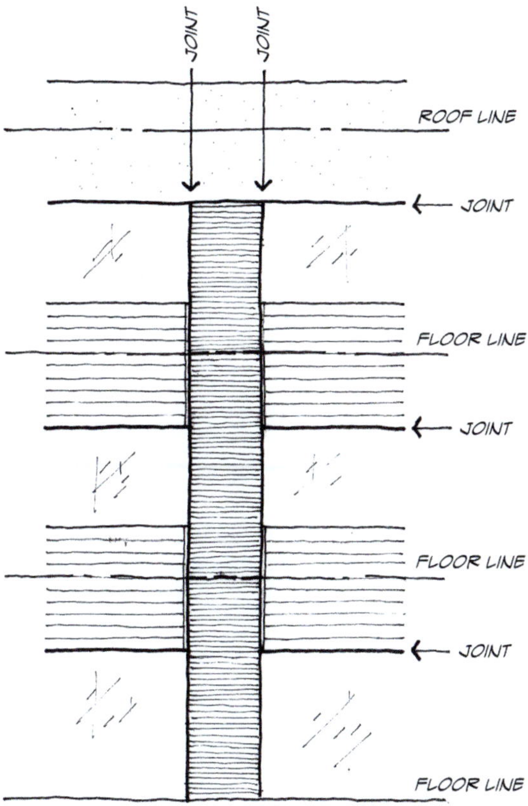

Figure 8.19 Building Elevation Showing Typical Joint Locations

continuous storefront ribbon isn't practical, owing to the architectural design of the building, locate horizontal joints in the cladding at the floor line to accommodate the anticipated expansion and contraction.

In the case of aluminum curtain wall systems, specifically designed thermal expansion points are incorporated into the system by the manufacturer, as shown in Figure 8.20. These joints occur behind the visible trim cover, so that they are all but invisible to the casual observer. In many of these systems, there are oversized glazing channels built in to allow for minor deformation of the aluminum system without the glass being broken.

8.3.2 Expansion and Contraction Related to Moisture

Some materials shrink after they are installed, and others expand. Although many cladding materials expand and contract in cycles for their usable life, there are a couple of notable exceptions. For instance, a nonreversible volumetric change occurs in concrete (contraction) and clay brick (expansion). Clay masonry will expand as it absorbs moisture and continue to do so until **saturated**, whereas concrete will contract as it

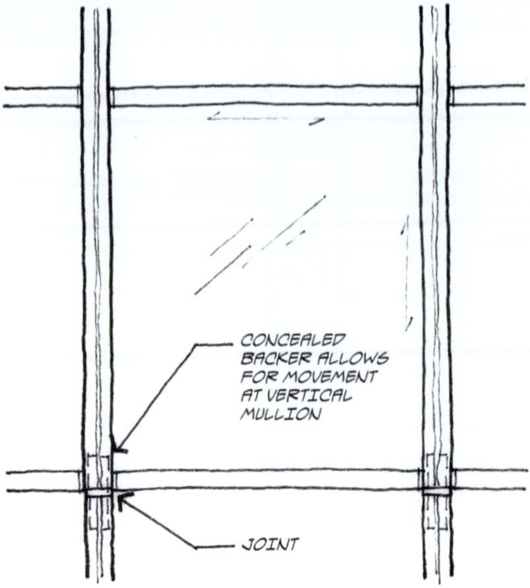

Figure 8.20 Window Elevation Showing Expansion Joints

cures, never returning to its original volume. Joint details to accommodate this type of movement at panel or section ends can be handled in much the same way as for thermal movement, but must consider the direction of dimensional change. Figure 8.21 shows a simple detail for a brick **ledger** in multistory construction that allows for thermal movement and brick growth. In a reinforced concrete frame, the joint size should consider column shortening due to shrinkage, brick lengthening from moisture, as well as anticipated thermal movement. The joint will likely be at its largest the day it is built and then become more of a thermal joint once the concrete and brick have come to a steady size.

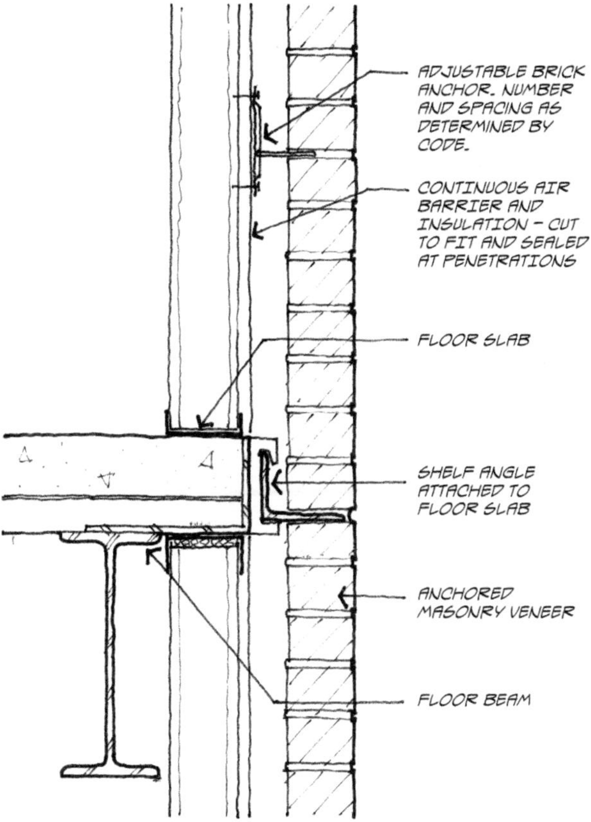

Figure 8.21 Brick Ledger Detail

8.3.3 Drift Movement

Drift is horizontal movement between stories in a building. As one floor moves relative to the other, it creates a **shearing** action that tends to make rectangular shapes into trapezoids (Figure 8.22). This movement can cause windows and cladding to crush. However, if we understand the problem, we can design for it.

Options to detail for drift include continuous horizontal window bands (Figure 8.23) and cladding panels that are offset and can slide in front of or behind each other (Figure 8.24). Corners are most difficult, as we have one portion tipping outward, while the perpendicular direction is trying to slide (Figure 8.25A). Effective ways to address this include a corner panel that covers the **spandrel** panels. This allows the spandrel panel to slide behind the corner cover, as shown in Figure 8.25B. Large joints can be used, but are visually unappealing and lead to water leaks. The key is to elegantly accommodate the building drift with simple, well-thought-out methods.

The information provided in Table 8.1 allows us to conceptualize the required drift movements between different cladding components. However, actual drift criteria will be provided by the structural engineer on the project, often resulting in a less restrictive design.

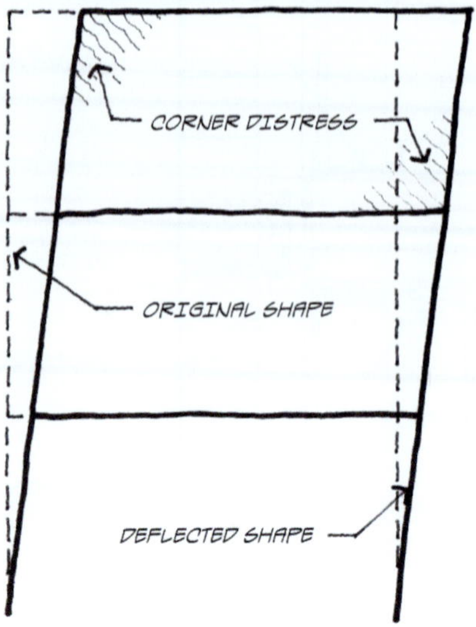

Figure 8.22 Effect of Drift Movements

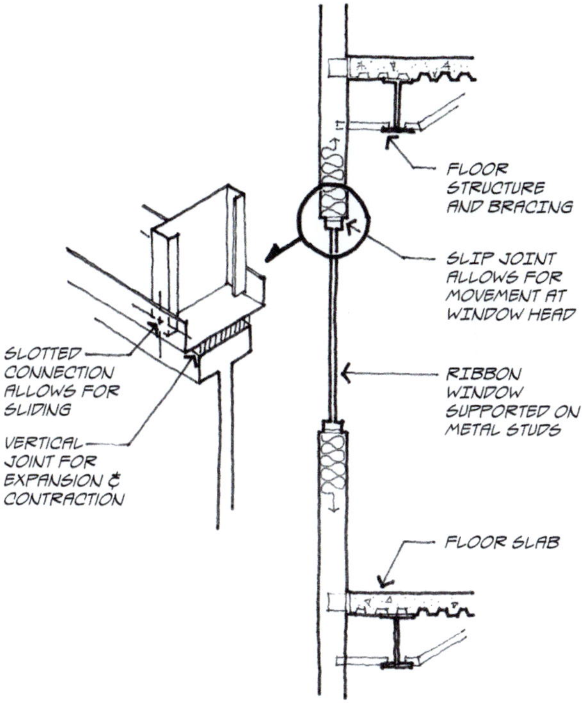

Figure 8.23 Window Joint Allowing Sliding between Stories

8.4 CONNECTION TO STRUCTURE

Coming full circle, we now consider how we connect the cladding to the primary structure. For lighter systems, the connections are made to the slab edge without much reinforcing, as shown in Figure 8.26. For heavier systems, such as precast concrete and stone, the connections are developed deeper into the slab–beam system, as illustrated in Figure 8.27.

8.4.1 Curtain Wall

Curtain wall connections fall in the lighter weight category. They are made to the concrete slab edge, as in Figure 8.26. A key consideration is the inclusion of a thermal break. This prevents the aluminum mullions from transmitting heat to the primary structure.

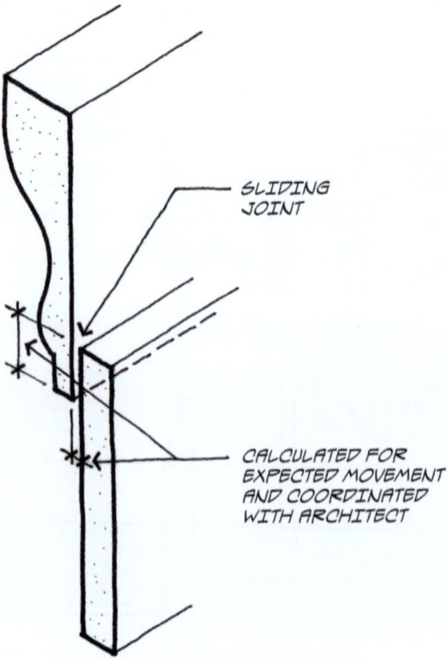

Figure 8.24 Offset Cladding Panels Allowing Movement

8.4.2 Stud

Light gauge metal studs are common and relatively simple to install. They usually sit on or hang from the slab, as seen in Figure 8.26. It is important to have a slip connection at the top when the stud runs full height, to allow the floors to deflect differently and not pick up high loads in the studs.

8.4.3 Precast and Stone Panels

Precast and stone panels impart much larger loads to the structure than curtain wall. These are often carried by heavier **embeds** in concrete slabs, such as that in Figure 8.27 for slab over metal deck in steel structures.

Some design considerations are worth noting with heavier panels. Where the spandrel panel (horizontal) is as long as the bay, it is best to impart the dead loads to the structure in two places near the column. This reduces the deflections in the spandrel beam and eliminates load sharing between the panel and floor beam. Often a precast or stone panel is stiffer

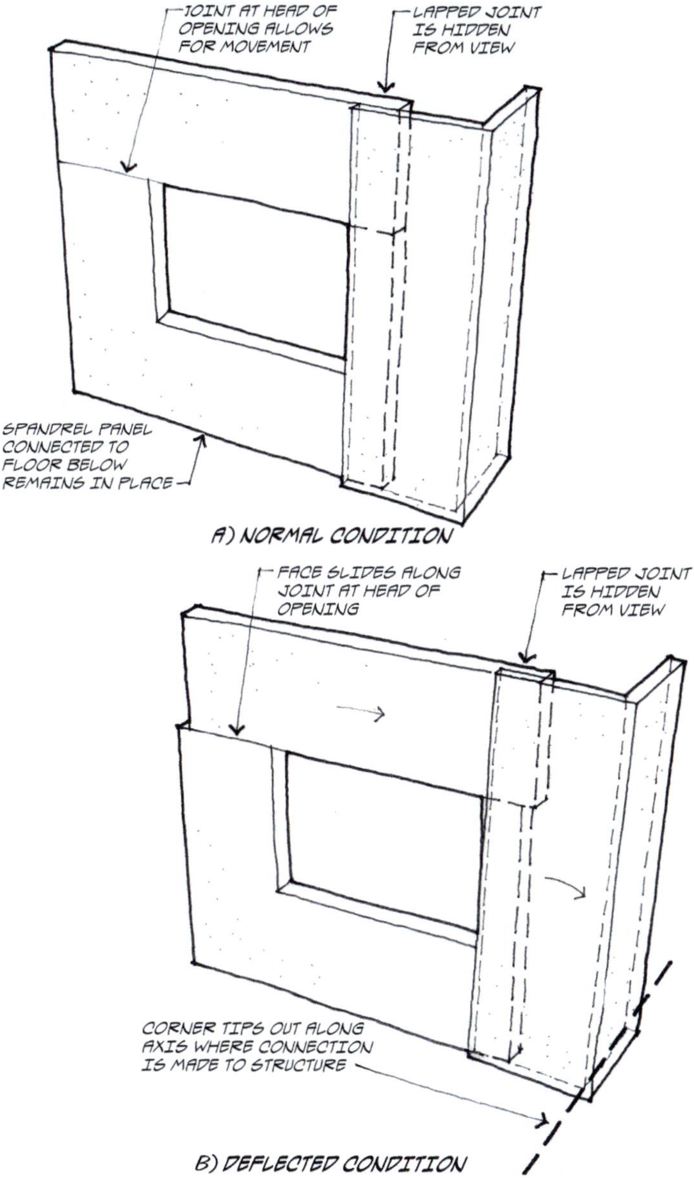

Figure 8.25 Corner Condition Showing Slip and Tip

Table 8.1 Drift Limits for Varying Construction Types and Heights

Drift Limits			
Structural System	*Risk Category*		
	I or II	*III*	*IV*
Structures 4 stories or fewer, non-masonry, with interior walls & ceilings designed to accommodate drift	$0.025h_{sx}$	$0.020h_{sx}$	$0.015h_{sx}$
Masonry cantilever shear wall structures	$0.010h_{sx}$	$0.010h_{sx}$	$0.010h_{sx}$
Other masonry shear wall structures	$0.007h_{sx}$	$0.007h_{sx}$	$0.007h_{sx}$
All other structures	$0.020h_{sx}$	$0.015h_{sx}$	$0.010h_{sx}$

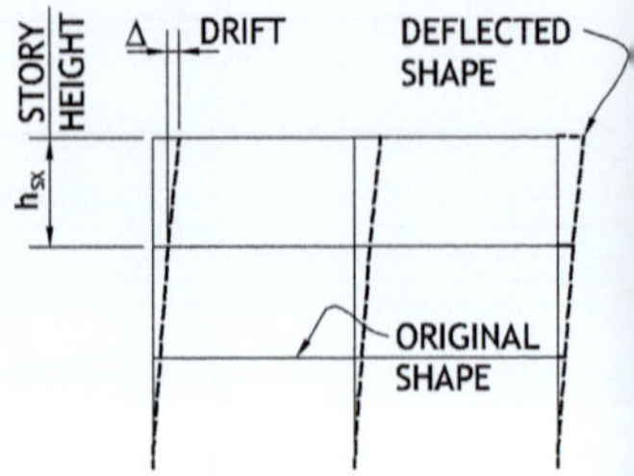

Limit Criteria	*Allowable Deflection Values for Various Criteria & Lengths*						
	Story Height (ft)						
	8	*10*	*12*	*14*	*15*	*16*	*20*
	Allowable Drift Δ_a (in)						
$0.007h_{sx}$	0.67	0.84	1.01	1.18	1.26	1.34	1.68
$0.010h_{sx}$	0.96	1.20	1.44	1.68	1.80	1.92	2.40
$0.015h_{sx}$	1.44	1.80	2.16	2.52	2.70	2.88	3.60
$0.020h_{sx}$	1.92	2.40	2.88	3.36	3.60	3.84	4.80
$0.025h_{sx}$	2.40	3.00	3.60	4.20	4.50	4.80	6.00

Note: h_{sx} = Story height under level being considered; don't forget to convert to inches or mm

Source: ASCE 2010

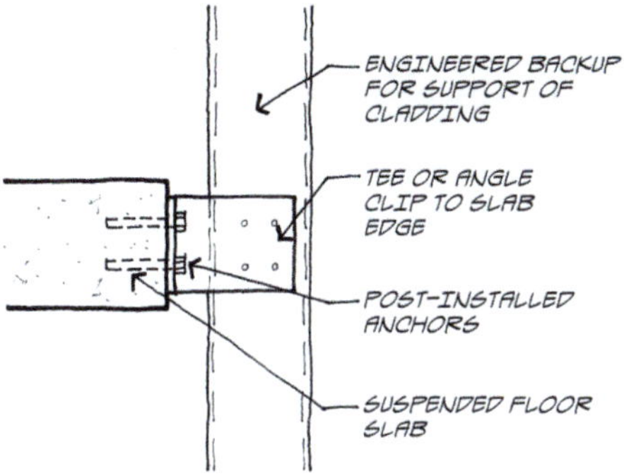

Figure 8.26 Light Cladding Connection to the Slab Edge

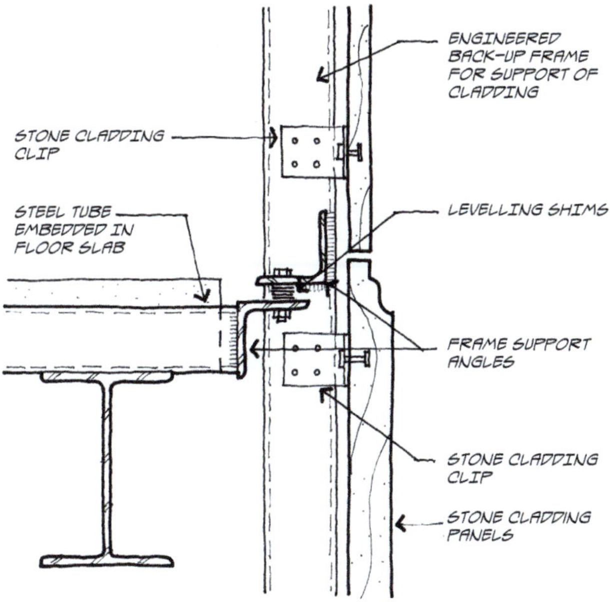

Figure 8.27 Heavy Cladding Connection to Slab

than the floor beam and, if not detailed well, can actually pick up floor load through the gravity connections. The panel can then be laterally braced at numerous locations along its length.

Similarly, for column covers, it is prudent to impart cladding loads near the floor lines, as seen in Figure 8.28. This helps reduce any bending **moments** in the columns, which can require a substantial increase in materials.

8.4.4 Connection Tolerance

Industry standard construction tolerances for the primary structure are typically far more permissive than those for cladding. This creates a need to provide adjustability in the cladding to allow for structural connections. For example, the first 20 stories in a steel building may be out of plumb at a slope of 1 in 500, with a maximum of 2 in (50 mm). Over 50 ft (15.24 m), this means the column and associated framing could be 1.2 in (30.5 mm) from where we think it is. As you detail this interface, leave room between the back of cladding and structure, as illustrated in Figure 8.29, for when

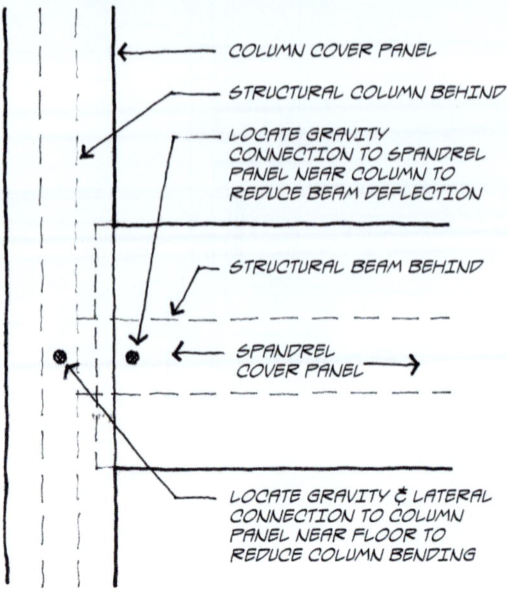

Figure 8.28 Preferred Connection Locations for Heavy Cladding

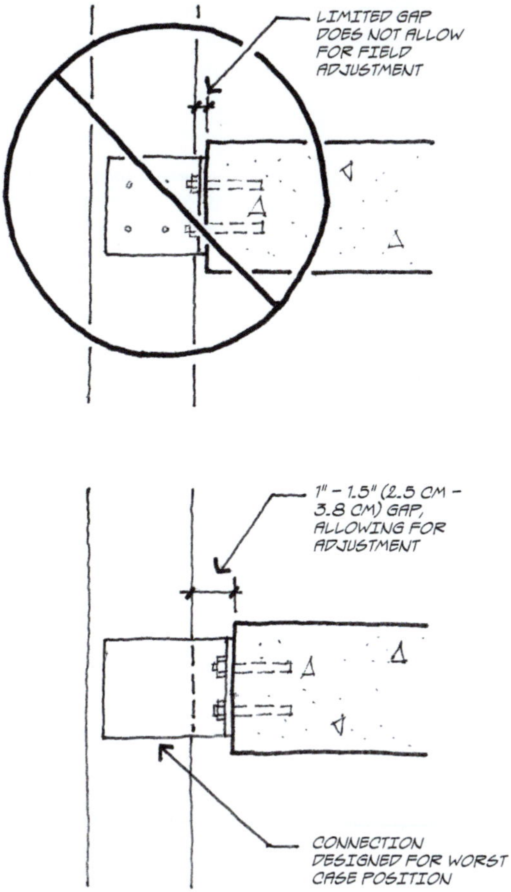

Figure 8.29 Slab Edge Detail Showing Necessary Clearance

things aren't built perfectly plumb, level, and square (in other words, all the time).

8.5 SUMMARY

As an architect leading the cladding design, or a contractor responsible for building it, get the discussion started early. Make sure your team is thinking about this fundamental piece of the building. Get everyone together, discuss the issues, and start fleshing out some simple details that will guide the full development of the cladding system. Where necessary, engage specialty help. There are people who do this all the time.

NOTES

1. ASTM. *Standard Specification for Heat-Strengthened and Fully Tempered Flat Glass*, ASTM C1048–12e1 (West Conshohocken, PA: ASTM International, 2012).
2. ASTM. *Standard Specification for Laminated Architectural Flat Glass*, ASTM C1172–14 (West Conshohocken, PA: ASTM International: 2014).
3. ASTM. *Standard Practice for Determining Load Resistance of Glass in Buildings*, ASTM E1300–12ae1 (West Conshohocken, PA: ASTM International, 2012).
4. ASTM. *Standard Specification for Nonstructural Steel Framing Members*, ASTM C645–14e1 (West Conshohocken, PA: ASTM International, 2014).
5. ASTM. *Standard Specification for Installation of Steel Framing Members to Receive Screw-Attached Gypsum Panel Products*, ASTM C754–15 (West Conshohocken, PA: ASTM International, 2015).
6. ASTM. *Standard Test Method for Structural Performance of Exterior Windows, Doors, Skylights and Curtain Walls by Uniform Static Air Pressure Difference*, ASTM E330/E330M-14 (West Conshohocken, PA: ASTM International, 2014).
7. ASTM. *Standard Test Method for Determining Rate of Air Leakage Through Exterior Windows, Curtain Walls, and Doors Under Specified Pressure Differences Across the Specimen*, ASTM E283–04 (West Conshohocken, PA: ASTM International, 2012).
8. ASTM. *Standard Test Method for Water Penetration of Exterior Windows, Skylights, Doors, and Curtain Walls by Uniform Static Air Pressure Difference*, ASTM E331–00 (West Conshohocken, PA: ASTM International, 2009).
9. AAMA. *Standard Test Method for Water Penetration of Windows, Curtain Walls and Doors Using Dynamic Pressure*, AAMA 501.1–05 (Schaumburg, IL: American Architectural Manufacturers Association, 2005).
10. AAMA. *Recommended Static Test Method for Evaluating Curtain Wall and Storefront Systems Subjected to Seismic and Wind Induced Interstory Drifts*, AAMA 501.4–09 (Schaumburg, IL: American Architectural Manufacturers Association, 2009).
11. AAMA. *Recommended Dynamic Test Method For Determining The Seismic Drift Causing Glass Fallout From A Wall System*, AAMA 501. 6–09 (Schaumburg, IL: American Architectural Manufacturers Association, 2009).
12. AAMA. *Test Method for Thermal Cycling of Exterior Walls*, AAMA 501.5–07 (Schaumburg, IL: American Architectural Manufacturers Association, 2007).
13. ASTM. *Standard Specification for Glazed Brick (Single Fired, Brick Units)*, ASTM C1405–16 (West Conshohocken, PA: ASTM International, 2016).
14. ASTM. *Standard Specification for Thin Veneer Brick Units Made From Clay or Shale*, ASTM C1088–14 (West Conshohocken, PA: ASTM International, 2014).
15. ASTM. *Standard Specification for Architectural Cast Stone*, ASTM C1364–16 (West Conshohocken, PA: ASTM International, 2016).

Fire Design

Chapter 9

Kevin J. LaMalva

The protection of structures from fire is critical, but remains one of the most misunderstood areas of building design. Typically, we employ **standard fire resistance** design (Option 1) to protect structures from fire. This method relies on an empirical indexing system and does not involve structural engineering principles. Alternatively, we may adopt a **structural fire engineering** approach (Option 2) that requires explicit evaluation of the structure's ability to endure fire, similar to how other design loads are treated in structural engineering practice.

We begin this chapter by exploring the reasons that we insulate structures against fire, and the effect of fire on structures. We will then examine the two structural fire protection design options, with practical guidance on their use. Lastly, we will take a glimpse into the future of structural fire protection.

9.1 WHY INSULATE STRUCTURES?

The frequency of uncontrolled fire in an engineered building is low, but the consequences can be severe. When structures are heated by fire, they experience thermal effects that are not contemplated by conventional structural engineering design. Under these conditions, the functionality of building safety systems assumes the structural system will remain stable. In light of this, we should ask the question: "Do building codes require that structures be engineered to perform adequately under fire?" To the surprise of many, the answer is actually no. Instead, structures are usually required to be insulated to reduce their heating, with the hope of the fire risk being mitigated to an acceptable level. This is unlike our consideration of hurricanes and earthquakes, in which structures are engineered to withstand these natural phenomena.

Fire sprinkler systems significantly reduce the probability of uncontrolled fire in buildings. However, we must appreciate the fact that these systems are generally not effective against large fires, such as one due to an explosion.[1] Fire sprinkler systems are only designed for a limited number of contiguous sprinklers (e.g., five sprinkler heads) to activate and spray water onto a single small fire. Moreover, the effectiveness of a system drops precipitously when more than 10 sprinkler heads are activated, as this begins to tax the water delivery.[2] Fire sprinkler systems can also be rendered ineffective under the following possible cases: water supply valves are accidently left shut (e.g., after maintenance); the system has not been properly maintained (e.g., pipe blockage); there are damaged mechanical components; and other cases. In these rare instances, structural fire protection is relied upon.

During an uncontrolled fire event, we expect a safe evacuation of building occupants and protection of those who take shelter in designated building refuge areas. Accordingly, we must appreciate that the total evacuation time increases as the height of the building increases, as occupants must traverse stairways from upper floors to reach the ground level. However, building codes do not require more structural insulation as the height of the building increases (e.g., from 30 stories to 200 stories). Building codes limit the distance that occupants must travel to reach exit stairways, but they do not regulate the total evacuation time. Also, the presence of designated refuge areas is more common in tall buildings, and they are expected to remain safe for a period of time during and after a fire event. These performance expectations are predicated on the assumption that the structural systems supporting evacuation routes and refuge areas in buildings remain stable.

Beyond occupant life safety, we may also expect that buildings possess a certain level of resiliency when exposed to an uncontrolled fire. However, building codes do not require confirmation of this, but prescribe certain levels of insulation to mitigate the risk.

9.2 WHAT IS "FIREPROOFING"?

"Fireproofing" is a term commonly used to describe insulation used for structural fire protection. This term is actually a misnomer, as applied insulation is only able to slow the heating of a structure, not prevent its heating during a fire. There are various methods to insulate structural elements, as described below.

For steel construction, the commonest type of insulation is spray-applied fire-resistive material (SFRM). SFRM adheres to the surface of steel and is an effective insulator owing to its low conductivity and chemical reactions under fire that absorb heat. For this case, it is important that the steel surface is prepared to prevent premature delamination. Other types of steel insulation include intumescent and board products. Intumescent paint is more expensive than SFRM, but offers better aesthetics. The paint expands in thickness and forms an insulating char layer when heated. Fire-resistive boards are fastened to structural elements in order to fully encapsulate them. Lastly, the use of a membrane is a protection method that involves hanging a fire-resistive suspended ceiling below the structural elements, which serves as a heat shield during **fire exposure**.

For concrete construction, the commonest method of fire protection is to specify a certain thickness of concrete cover to insulate the steel reinforcement. If concrete cover alone does not suffice, spray-applied or board assemblies could also be used. Heavy timber construction relies on sacrificial wood charring for fire protection. In this case, the outer layer of the heavy timber member chars and forms an insulating layer, helping to slow further degradation. Also, the reliance on wood charring can be supplemented with spray-applied or board assemblies, similar to concrete construction.

9.3 EFFECTS OF FIRE ON STRUCTURES

There are two primary types of fire encountered within building spaces: enclosure and localized. As shown in Figure 9.1a, an enclosure fire contains hot gases by walls and ceilings, causing the formation of distinct upper and lower gas layers (e.g., fire within an office). The upper layer descends downward as the fire progresses. In a localized fire, hot gases are contained within a plume (e.g., fire within a building atrium that is more open). Enclosure fires heat structures more uniformly compared with localized fires. In both cases, structures are heated by the hot gases and radiation from the fire itself. If the fire is intense enough, flames may also directly heat structural elements.

When the temperature of steel exceeds approximately 600°F (316°C), it begins to lose significant strength and stiffness rapidly, which can result in deflections that far exceed those contemplated by conventional structural

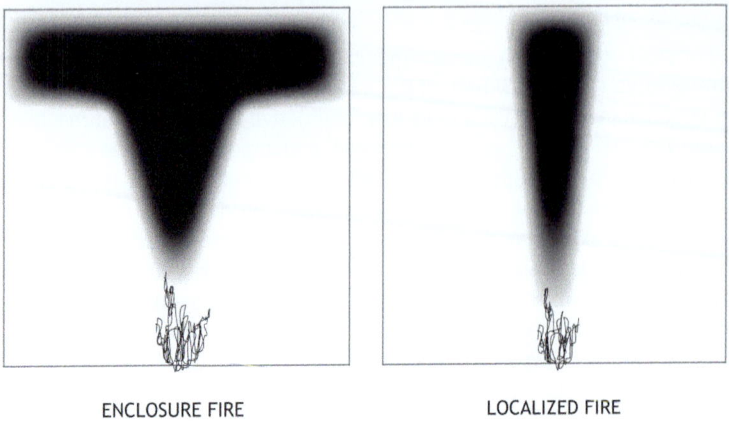

Figure 9.1 (a) Enclosure Fire vs. (b) Localized Fire

design. These large deflections induce added forces and rotations at structural connections. To make matters worse, the heating of steel members causes them to expand longitudinally, which increases connection forces. Similar to steel, concrete loses strength and expands under fire exposure. Concrete can experience spalling, which occurs when trapped moisture in the concrete is vaporized, leading to explosive dislodgment of concrete. Heavy timber construction does not expand under fire exposure, but it undergoes section loss as a char layer forms. This reduces the effective cross section that can resist structural loads.

If structural systems are not robust enough to accommodate **fire effects**, they may experience premature local or global collapse. The industry recognizes two design approaches for fire protection: *standard fire resistance design* and *structural fire engineering.*

9.4 OPTION 1: STANDARD FIRE RESISTANCE DESIGN

Since the early twentieth century, building codes have primarily adopted standard fire resistance design to mitigate the effects of fire on structures. Standard fire resistance design is based on small-scale component testing and relies almost exclusively on protective insulation for structural fire protection. This method has not significantly changed since its very early inception, and designers and academics routinely criticize this approach for its lack of technical basis and indeterminate performance expectations. The reliance of this method on protective insulation is analogous to constructing a tall obstruction in front of a building in order to protect it from a hurricane, instead of analyzing how its structure actually performs under wind loads.

9.4.1 Fire Resistance Requirements

As shown in Table 9.1, the required level of fire resistance is based upon the construction type of the building, which ranges from Type I-A (most restrictive) to Type V-B (least restrictive). For each construction type, building codes prescribe the required fire resistance of the primary and secondary structural system, designated in hours. For a given building occupancy classification (e.g., business), building codes limit the allowable height and area of the building as a function of the construction type. For instance, high-rise buildings are usually required to be Type I. In some cases, the required levels of fire resistance can be reduced if the building has a fire sprinkler system and/or provides exceptional frontage for fire department access.

Table 9.1 Fire Resistance Requirements

Structural Element	*Type I (Noncombustible Construction)*		*Type II (Noncombustible Construction)*		*Type III (Noncombustible Exterior Walls)*		*Type IV (Heavy Timber)*	*Type V (Combustible Construction)*	
	A	*B*	*A*	*B*	*A*	*B*	*N/A*	*A*	*B*
Primary structural system (columns, members connecting directly to columns, and lateral bracing)	3	2	1	0	1	0	Specific requirements	1	0
Secondary structural system (floors)	2	2	1	0	1	0	Specific requirements	1	0
Secondary structural system (roofs)	1–½	1	1	0	1	0	Specific requirements	1	0

Note: "A" and "B" designations are each a subclass of the main construction type

Source: *International Building Code*, 2015, International Code Council[3]

9.4.2 What Is Fire Resistance?

It is often misunderstood that 2 hours of fire resistance means that a structure would survive a fire for 2 hours without failure. However, it is possible that a "2-hour" structure could survive only 30 minutes, or maybe 36 hours. If this sounds bizarre, please let me explain what "2 hours" of fire resistance actually means.

Standard fire resistance design is based on small-scale furnace testing of structural components. As shown in Figure 9.2, structural systems are decomposed into individual components that are tested in isolation. Accordingly, this testing does not include structural connections or any representation of the structural system beyond each component. So that a component can be qualified as having a certain level of fire resistance, a furnace is used to conduct the testing in accordance with either ASTM E 119[4] or UL 263,[5] which are nearly identical test standards. With the motivation of selling fire-resistant insulation, fire protection manufacturers pay a significant fee to have a component tested with their insulation product used, in order to obtain fire resistance qualification for a specific assembly.

Owing to the small size of test furnaces, the span of beams and floors is limited to 17 ft or less. For tests of floors, hydraulic actuators are used to apply loading to represent the design live load. The furnace produces a very intense fire exposure from below, and the temperature of the mock-up assembly is measured. When either the assembly temperature exceeds predetermined limits, or the assembly has deflected significantly (i.e., to a point at which the test operators fear for the condition of the furnace apparatus), the evaluators stop the test, record the time, and establish the qualified fire resistance period.

9.4.3 Specifying Fire Resistance

Once you know what level of fire resistance is required, you can specify a corresponding fire resistance rated assembly. The UL *Fire Resistance Directory* is a large collection of qualified assemblies (also known as listed assemblies) organized by assembly type (e.g., columns, beams).[6] Each assembly has a unique UL design number that you can reference in your specifications.

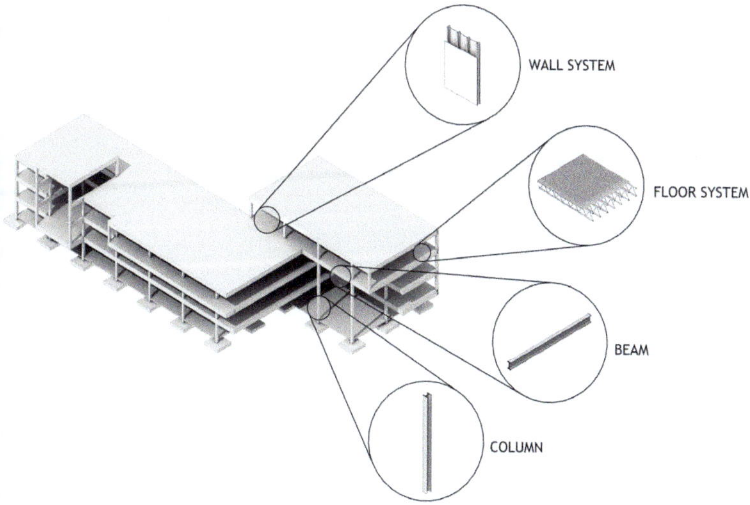

Figure 9.2 Structural System Decomposition for Furnace Testing

UL listings contain very detailed construction requirements, as each listing is essentially a memorialization of what was tested in a furnace, and you may not deviate from these instructions unless there are specific exceptions written into the listing. Naturally, you may feel quite restricted by this, and architects often flip through the pages of the UL *Directory* endlessly without finding an assembly that suits all of their design objectives.

In addition to the UL *Directory*, building codes often provide empirical equations that can be used to determine the fire resistance of assemblies composed of conventional building materials (e.g., concrete beam). These equations are generally derived by interpolation between available furnace test data. Also, you may reference the ASCE/SEI/SFPE 29 standard that provides similar interpolations.[7] Beyond the UL *Directory* and related empirical equations, you may be able to demonstrate analytically that a proposed assembly is equivalent to one that has been tested. In the relatively limited number of cases where this approach can be applied, the analysis would involve predicting or simulating the results of a furnace test. The "Rules of Harmathy" are often used to demonstrate such equivalence.[8]

9.4.4 The "Restrained vs. Unrestrained" Paradox

Starting in the 1970s, ASTM introduced the "restrained vs. unrestrained" concept. This concept is based upon the observation that assemblies confined against thermal expansion during furnace testing generally achieve higher fire resistance compared with those that are free to expand.[9] For instance, a steel beam and concrete slab assembly would be considered "restrained" if it bore directly against the side of the furnace at the outset of the test. Many listings in the UL *Directory* permit less insulation to achieve a certain fire resistance rating if the designer can demonstrate that the assembly will be "restrained" when it is actually constructed as part of a structural system.

The "restrained vs. unrestrained" concept is flawed and often misinterpreted. As there are significant insulation savings to be had, it is often asked if the in situ condition of a listed assembly could be considered "restrained." The UL *Directory* and other similar standards state that whether an assembly is "restrained" or "unrestrained" must be determined by the judgment of the designer.

Considering the example above, the steel beam and concrete slab would be restrained equally during a furnace test. In actual building construction, the beam would typically have less resistance to thermal expansion compared with the slab, resulting in differential longitudinal movement under fire exposure. Also, the UL *Directory* states that a furnace provides approximately 850 million lb-in of flexural restraint, which is an extremely high stiffness that actual structural systems can't provide. Consequently, there continues to be ongoing confusion and debate concerning this concept.

Until the industry clears up the restrained vs. unrestrained paradox, designers may choose to take a conservative approach when classifying restraint conditions. Notably, IBC Section C703.2.3 states that in-place conditions should be considered unrestrained unless structural documentation is provided that demonstrates a restrained condition in actual construction. In all cases, the authority having jurisdiction may be consulted as to the proper interpretation for a given project.

9.5 OPTION 2: STRUCTURAL FIRE ENGINEERING

As an alternative to standard fire resistance design, structural fire engineering explicitly evaluates the demand and capacity of structural systems under fire loading in a similar manner as other design loads are treated in structural engineering practice. Within this framework, the demand on a structural system under fire loading can be reduced by means of rationally allocated structural insulation, control of fuel loads, and other fire exposure mitigation techniques. Also, the capacity of a structural system to endure fire effects can be increased by means of specific member sizing, connection detailing, and other measures to enhance structural robustness with respect to explicit performance objectives.

The 2016 edition of ASCE/SEI 7 (*Minimum Design Loads and Associated Criteria for Buildings and Other Structures*) permits the use of this approach in accordance with the new Appendix E: *Performance-Based Design Procedures for Fire Effects on Structures.*[10] This pertains to the structural system only, excluding nonstructural assemblies such as fire barrier walls that are governed by their flame passage resistance, which is best qualified using Option 1. Notably, the inclusion of Appendix E in ASCE/SEI 7 marks the *first time that fire effects are considered as an explicit design load condition in a U.S. structural engineering standard.*[11]

9.5.1 Performance Objectives

ASCE/SEI 7 Appendix E specifies mandatory performance objectives that pertain to occupant life safety. First, the structural system must allow for a safe and complete evacuation of building occupants to a public right of way in the event of an uncontrolled fire. In order to demonstrate this, the designer must employ an "ASET vs. RSET" analysis. The available safe egress time (ASET) can be determined by analysis of the endurance of the structural system to fire exposure. The required safe egress time (RSET) can be determined by analysis of the time it would take occupants to evacuate, perhaps using people movement computer software. Second, the structural system that supports refuge areas must remain stable during and after an uncontrolled fire exposure. The evaluation of RSET and identification of refuge areas are best performed by a fire protection engineer, and the evaluation of ASET is best performed by a structural engineer.

Appendix E also provides guidance on discretionary performance objectives that may be agreed upon and enacted by project stakeholders (including the building authority), such as enhanced building resiliency. This agreement should be memorialized in a design brief document before the onset of the structural fire engineering design.

9.5.2 Thermal Response (the Demand)

The first step in a structural fire engineering analysis is to estimate the heating of structural elements. ASCE/SEI 7 Appendix E requires that structural design fires be analyzed, which are those fires that are not controlled by fire sprinkler systems or manual intervention. SFPE S.01, *Engineering Standard on Calculating Fire Exposures to Structures*, provides design equations to estimate the heat produced from uncontrolled fire.[12] In cases where these equations are not applicable, computational fluid dynamics computer modeling could be used.

Once the heating from the structural design fire has been estimated, the temperatures experienced by the structural elements can be calculated using principles of **heat transfer**, in accordance with SFPE S.02: *Engineering Standard on Calculation Methods to Predict the Thermal Performance of Structural and Fire Resistive Assemblies.*[13] If the structural elements are insulated, there would be a significant delay in heating. In most cases, an insulated structural element can be idealized as two-dimensional and solved with either hand calculations or finite element software.

9.5.3 Structural Response (the Capacity)

Once we have determined the temperature histories of the structural elements, we can analyze the structural system's response. ASCE/SEI 7

Appendix E requires that all heated members be analyzed, as well as those unheated members that provide **thermal restraint**. For instance, a bridge with longitudinal expansion joints would not require us to analyze the unheated adjacent structures, granted that the expansion capacity of the joints exceeds the thermal expansion magnitude of the bridge under fire exposure.

In addition to potential forces induced by thermal restraint, we need to consider the effect of material strength and stiffness degradation, which may result in high deflections and deformations. To analyze these effects for steel structures, AISC 360 Appendix 4: *Structural Design for Fire Conditions* provides design equations that are similar to those for ambient design, but account for the degradation of steel's mechanical properties at elevated temperatures.[14] These equations are limited to single-member analyses and do not consider post-yield or post-buckling behaviors, which necessitate the use of finite element software.

During uncontrolled fire exposure, we expect that a structural system will undergo some level of permanent damage. Accordingly, ASCE/SEI 7 Appendix E allows for load redistribution and/or nonconventional sources of load carrying capacity to be relied upon. For instance, if the flange of a W-shape steel beam buckles locally, we may still be able to demonstrate that the structural system has adequate integrity to achieve the required performance objectives.

9.6 COMPARISON OF DESIGN OPTIONS

We have discussed the two design options permitted by the 2016 edition of ASCE/SEI 7, which are standard fire resistance design (Option 1) and structural fire engineering (Option 2). Let us examine the key differences between these two approaches.

9.6.1 Thermal Demand

Option 1 employs the same fire exposure for all furnace testing. As shown in Figure 9.3, the furnace fire exposure continually increases in temperature and lasts indefinitely. Conversely, Option 2 considers actual fire exposures, which typically increase in temperature to a limit and then decay as combustibles are consumed.

Option 1 tests assemblies within a furnace that is highly insulated, only allowing for heat dissipation through the top of the assembly to the ambient (e.g., from the top of a concrete slab). Real structures have heat

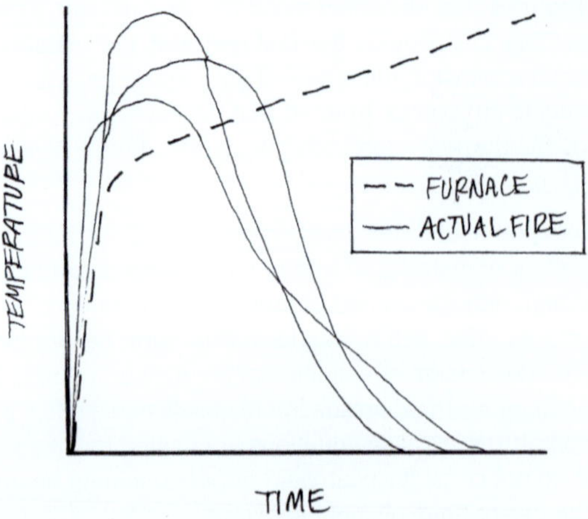

Figure 9.3 Comparison of Furnace and Actual Fire Exposures

sinks in all directions. Option 2 considers the possible heat dissipation to all adjacent structural elements.[15] For instance, Figure 9.4 shows a steel connection at the end of a heated beam, in which heat from the connection is dissipated to the concrete slab, as well as the adjacent column. Overall, there may be a significant difference between the level of heating demonstrated by Option 1 and what the structural system actually experiences during a fire.

9.6.2 Performance Expectation

Option 1 does not inform us about the actual performance of a structural system under fire exposure. As mentioned, there is widespread misconception about what "2 hours" of fire resistance actually means. For instance, consider the two steel assemblies shown in Figure 9.5. Based on its relative structural robustness, we would expect that the bottom assembly would perform better in a fire compared with the top assembly shown. However, if both of these assemblies are protected with 1 in (25 mm) of SFRM insulation, in accordance with UL Design No. N706, they would both be qualified for 2 hours of fire resistance.[16]

Option 1 does not consider the intended functionality of building egress features. For instance, Figure 9.6 shows three occupant evacuation scenarios within a high-rise building. This building has 16 stories, but

Figure 9.4 Thermal Modeling Showing Heat Dissipation from a Beam Connection

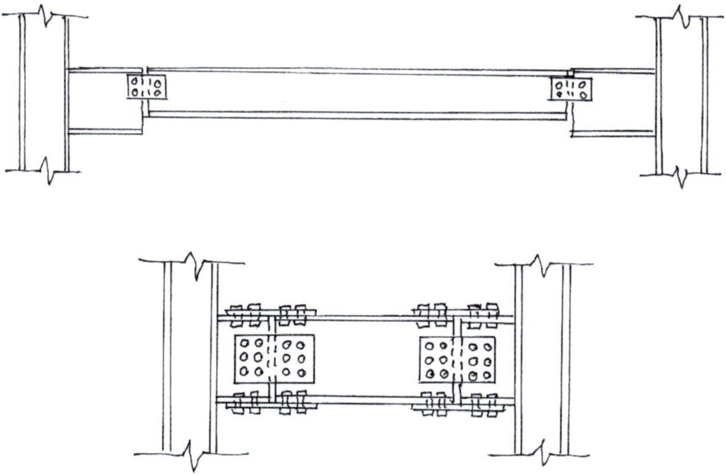

Figure 9.5 Hypothetical Steel Structures

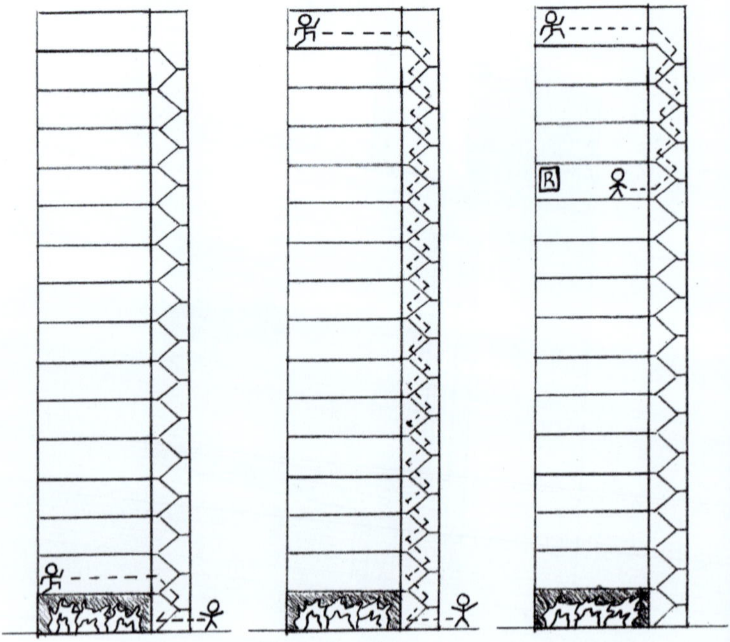

Figure 9.6 Hypothetical Evacuation Scenarios

could just as well be 200 stories tall. The first scenario shows a case in which a building occupant is able to evacuate the building quickly during a fire event. The second scenario is similar, except that the occupant must descend the entire height of the building via the exit stairway, significantly increasing evacuation time. In the third scenario, the building has a designated refuge floor, where an occupant is directed to take shelter. Considering these varying scenarios, we could identify a hierarchy of structural elements based on their criticality to occupant life safety. For instance, the structural members supporting the exit stairway shaft would be more critical than the secondary floor framing at the first-floor level. Accordingly, the level of insulation should be strategically varied throughout the building. Option 1 does not contemplate the configuration of the actual building. Conversely, Option 2 considers the hierarchy of structural elements and allows for strategic application of insulation to achieve the required performance objectives.

9.6.3 Design Freedom

As shown in Figure 9.7, Option 1 only allows you to influence the thermal "demand" on a structure by controlling its heating with insulation. Option 2 permits the designer to influence all variables on both sides of the demand and capacity comparison. This allows us to better optimize the design, based on aesthetics, economy, and functionality. Hence, Option 2 provides enhanced design freedom in situations where Option 1 is found to be overly restrictive or unsuitable for the application. For instance, Option 2 may be necessary as part of building code variances in order to demonstrate the adequacy of innovative and/or nonconventional architecture. Additionally, this approach allows for efficient analyses of nonconforming existing building construction (e.g., historic preservation) prior to costly rehabilitation.

9.6.4 Insulation

Figure 9.8 shows a hypothetical sampling of the building population that ranges across the horizontal axis from Type-V-B (least restrictive) to Type-I-A (most restrictive) construction. The buildings shown form a vertical bar chart, and their height symbolizes the actual amount of insulation required to meet minimum performance expectations. You may notice that, within each construction-type category, the actual amount of insulation required varies, even though Option 1 would consider these equal. This is because the robustness of the host structural system varies, and only Option 2 considers this. For instance, a building may have seismic connections that provide for enhanced ductility under fire exposure.

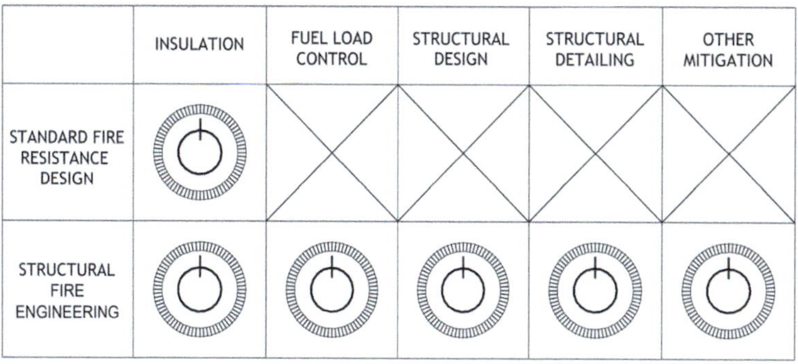

Figure 9.7 Controllable Design Variables

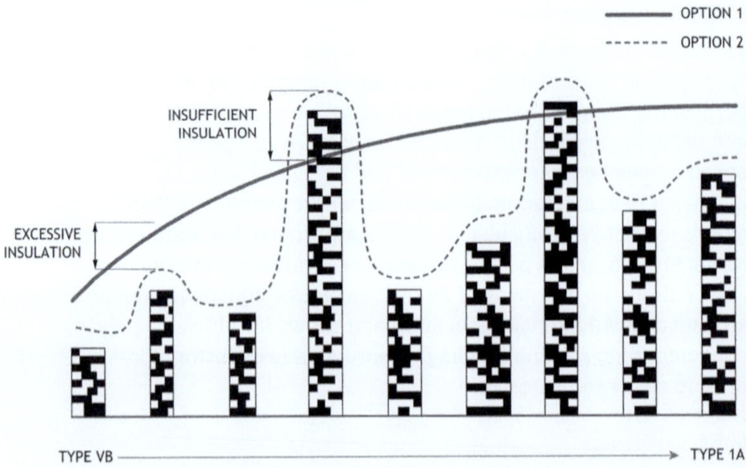

Figure 9.8 Required Level of Insulation

As the required level of insulation for Option 1 is based solely on the construction type of the building, there may be cases where the structural system is highly over-insulated and, perhaps, rare cases where insufficient insulation is provided. Consequently, Option 1 provides a safety margin that is nonuniform. Conversely, Option 2 provides a consistent safety margin, as the applied insulation is adapted to the characteristics of each structural system. This presents opportunities to save costs on insulation in many cases and better protect those structures that are particularly vulnerable to fire exposure in other cases.

9.7 PRACTICAL USE OF DESIGN OPTIONS

You may be thinking at this point, "Why would I ever want to use Option 1?" I hope that I have shed some light on this approach, but I certainly do not wish to discourage you from using it in practice. In fact, for the majority of buildings, Option 1 is adequate. The good practices of this approach promote high-quality building construction in terms of its resistance to fire exposure. For instance, imagine how well a gypsum-board encasement assembly would need to be constructed for it to survive 2 hours in a furnace. This empirical approach has a great historical track record for conventional buildings. Also, Option 1 requires far less engineering rigor compared with Option 2. In general, it is prudent for you to use Option 1 for conventional buildings, where the associated requirements do not impart undue hardship upon the project stakeholders or upon your design aspirations.

In other cases, the design team may seek relief from the often-restrictive requirements of Option 1 by adopting the Option 2 approach. This can address issues related to building aesthetics, functionality, construction economy, and justification of existing construction. Also, the building authority may require an Option 2 approach in cases where the building has a high consequence of failure (e.g., high-rise buildings) or specific potential threats (e.g., terrorism). In these cases, the added design costs associated with Option 2 may be justified, sometimes many times over. Knowledge of both approaches will assist you in handling structural fire protection issues as they arise, and will not let them derail your design objectives.

9.8 FUTURE VISION

Now that we have examined the current state of the industry, let us take a look into the future of structural fire protection. In terms of the Option 1 approach, I foresee that the hourly fire resistance ratings in the building codes will be changed to "fire resistance grades." For instance, "Fire Resistance Grades A, B, C, and D" could replace 1-hour, 2-hour, 3-hour, and 4-hour fire resistance ratings. This would help dispel major misconceptions about the significance of hourly ratings and reinforce, to designers and building authorities, that Option 1 is an empirical index, not a prediction of the actual survival time of a structure.

Although rare, severe structural damage and collapses of engineered buildings due to fire (e.g., One Meridian Plaza building,[17] World Trade Center (WTC) 1 and 2 buildings,[18] WTC 5 building,[19] WTC 7 building,[20] Windsor Tower,[21] Delft Technical University building,[22] and others) have highlighted the need for structural engineering participation in the evaluation of fire effects on structural systems. Moreover, the failure modes exhibited by these events could not have been anticipated by examination of furnace testing results (e.g., connection failures). Accordingly, I foresee that building codes will begin to incentivize the adoption of an Option 2 approach as we move forward. For instance, they may raise the required level of fire resistance for buildings that have a high consequence of failure, unless an Option 2 approach is adopted. Since the early twentieth century, the required levels of fire resistance have actually been relaxed significantly, owing primarily to reductions granted for buildings equipped with a fire sprinkler system. However, I foresee the industry rethinking this trade-off, as the actual intent of structural fire protection is to protect structural systems when active systems are rendered inoperative, not when they work.

The application of an Option 2 approach for building projects has merit and enormous potential. Hence, I foresee this design approach transitioning from a niche consulting service to more common application, with building authorities requiring this approach more frequently. Also, I foresee the Option 2 approach transitioning from a deterministic framework (i.e., analyze an explicit set of structural design fires) to a more mature probabilistic framework (i.e., design for a certain risk-informed fire exposure). This transition would further align structural fire engineering with other areas of structural engineering.

In closing, architects hold tremendous influence in the area of structural fire protection. Accordingly, I hope that this chapter has provided useful commentary on where the industry currently is, and where it may be going in the future. I also hope that this information will assist you in your ongoing pursuit to design a better building.

NOTES

1. NFPA. *Standard for the Installation of Sprinkler Systems*, NFPA 13 (Quincy, MA: National Fire Protection Association, 2016).
2. J.R. Hall. *U.S. Experience with Sprinklers and Other Automatic Fire Extinguishing Equipment* (Quincy, MA: National Fire Protection Association, 2009).
3. ICC. *International Building Code* (Washington, D.C.: International Code Council, 2015).
4. ASTM. *Standard Test Methods for Fire Tests of Building Construction and Materials*, ASTM E 119 (West Conshohocken, PA: ASTM International, 2016).
5. UL. *Standard for Fire Tests of Building Construction and Materials*, UL 263 (Fremont, CA: Underwriters Laboratories, 2015).
6. UL. *UL Fire Resistance Directory* (Fremont, CA: Underwriters Laboratories, 2015).
7. ASCE. *Standard Calculation Methods for Structural Fire Protection*, ASCE/SEI/SFPE 29 (Reston, VA: American Society of Civil Engineers, 2005).
8. T.Z. Harmathy. "Ten Rules of Fire Endurance Ratings," *Fire Technology* 1, no. 2 (May 1965): 93–102.
9. ASTM. Standard Test Methods for Fire Tests of Building Construction and Materials, ASTM E 119 (West Conshohocken, PA: ASTM International, 1971).
10. ASCE. *Minimum Design Loads and Associated Criteria for Building and Other Structures*, ASCE/SEI 7 (Reston, VA: American Society of Civil Engineers, 2016).
11. N. Post. "9/11 Blazes Debunk Code Assumptions in Open-Plan Offices," *Engineering News Record* (August 2013).
12. SFPE. *Engineering Standard on Calculating Fire Exposures to Structures*, SFPE S.01 (Gaithersburg,

MD: Society of Fire Protection Engineers, 2001).

13. SFPE. *Engineering Standard on Calculation Methods to Predict the Thermal Performance of Structural and Fire Resistive Assemblies*, SFPE S.02 (Gaithersburg, MD: Society of Fire Protection Engineers, 2015).
14. AISC. *Specification for Structural Steel Buildings*, AISC 360 (Chicago, IL: American Institute of Steel Construction, 2010.
15. K. LaMalva. "Thermal Response of Steel Structures to Fire: Test versus Field Conditions," *Journal of Fire Protection Engineering* 21, no. 4 (2011): 285–299.
16. UL. *UL Fire Resistance Directory*, UL Design No. N706 (Fremont, CA: Underwriters Laboratories, 2015).
17. H. Eisner and W. Manning. "One Meridian Plaza Fire," *Fire Engineering* 144, no. 8 (1991): 51–58.
18. J. Gross and T. McAllister. *Structural Fire Response and Probable Collapse Sequence of the World Trade Center Towers*, NIST NCSTAR 1–6 (Gaithersburg, MD: National Institute of Standards and Technology, 2005).
19. K. LaMalva, J. Barnett, and D. Dusenberry. "Failure Analysis of the World Trade Center 5 Building," *Journal of Fire Protection Engineering* 19, no. 4 (2009): 261–274.
20. T. McAllister. *Structural Response and Probable Collapse Sequence of World Trade Center Building 7*, NIST NCSTAR 1–9 (Gaithersburg, MD: National Institute of Standards and Technology, 2008).
21. I. Fletcher. "Performance of Concrete in Fire: Review of the State of the Art, with a Case Study of the Windsor Tower Fire," *Proceedings of the 4th International Workshop for Structures in Fire*, Vol. 2, 2006.
22. Brian Meacham, Haejun Park, Michael Engelhardt, Adam Kirk, Venkatesh Kodur, Ijsbrand van Straalen, Johan Maljaars, Kees van Weeren, René de Feijter, and Kees Both. "Fire and Collapse, Faculty of Architecture Building, Delft University of Technology: Data Collection and Preliminary Analysis," CMMI Research and Innovation Conference, National Science Foundation, 2010.

Quality and Inspection

Chapter 10

William A. Komlos

"**Quality** is conformance to requirements."[1] This simple statement is the focus for the quality control and quality assurance efforts made throughout construction today. Once the design requirements are defined, inspection actions begin. The **destructive tests** and **nondestructive examinations** described in this chapter are part of the inspection actions taken that determine the quality of products and services brought together in structural projects.

Architects, with the support of the various specialized engineering teams, establish the project requirements when designing structures. Local building jurisdictions often impose additional tests and **special inspections** of various critical fabrication and installation operations. These inspections help to establish if the construction is in compliance with the building **code**. At project's end, the fabricators and installers will have certified in writing that their products and services are in conformance with the contract documents. Certificates of Compliance, legal statements imposed by the International Building Code (IBC)[2] and the regulations of various local jurisdictions,[3] bind the signers to the full weight of the law should their certified affirmations later prove false. The strength of that certification is assured by inspection.

The engineer is expected to determine the extent of verification testing and the **acceptance requirements**. Incorporation of appropriate standards and codes into the project specification is the first step to ensuring that vendors are meeting industry-recognized quality standards. In the United States, the American Society for Testing and Materials (ASTM) establishes testing methods for materials and manufactured products. For concrete structures, the American Concrete Institute (ACI) guides the design team and contractor on which ASTM tests will be required to measure concrete quality. For steel connections, the Research Council on Structural Connections has developed various functional tests that verify fastener assemblies will tighten high-strength bolted connections as designed. The American Welding Society (AWS) establishes welding requirements and nondestructive examination (NDE) methods to evaluate the quality of welds.

10.1 TESTING AND EXAMINATION

Testing methods generally destroy the test specimen to verify a physical attribute—think concrete cylinders crushed in a lab. NDE methods cause no damage to the item under inspection. For welds, NDE is routinely performed, with various methods being used, each providing its own unique indication of weld quality within the limits of the method.

10.1.1 Concrete

The ACI establishes the minimum design and quality requirements for structural concrete construction. ACI 318, *Building Code Requirements for Structural Concrete*,[4] specifies the tests required to evaluate concrete materials and mix compositions. Physical tests of cast concrete are identified so they can be traced to the actual lift or element in the completed structure. Test methods are defined by ASTM standards. The slump test measures the initial workability of a concrete mix. Mixed concrete is poured into a specific slump test, or Abrams, cone. The cone is filled in three lifts and rodded 25 times after each lift. After the final lift has been rodded, the cone is carefully removed, and the distance the wet sample slumps is measured. How the sample slumps is also evaluated to offer further information to the examiner regarding the cause of a failed test. Automated processes are available to measure slump during transit and at time of delivery. Water or superplasticizers can be added to adjust the slump prior to casting.

After curing for a specified time (usually 14 and 28 days), axial compression tests are performed on field-cast, laboratory-cured concrete cylinders to determine if the specified minimum design compression strength has been achieved. These tests are performed by compression of the cylinders in load-testing machines until failure. A specimen's ultimate compressive strength is calculated by the recorded load placed on the specimen being divided by its cross-sectional area. The Schmidt hammer test can nondestructively evaluate concrete strength. The test incorporates a specialized, spring-loaded hammer. The equipment measures how quickly the ball bounces back from hitting the concrete. The ball bounces off harder materials faster than off softer materials. Hardness numbers are derived based upon the relative speed with which the ball bounces back into the hammer. The hardness numbers are then equated to various strengths published in hardness conversion tables. This family of hardness testing is commonly called time-of-flight rebound tests. Similar tools are used to measure the hardness and strength of steel.

ACI 117[5] also specifies tolerances for the placement of reinforcing steel. The size and layout of reinforcing steel in the concrete are critical to the structural performance. For this reason, reinforcing bar location, lap lengths, spacing, and clearance to forms are measured and visually examined prior to the concrete being cast.

10.1.2 Timber

The American Wood Council (AWC) maintains the *National Design Specification for Wood Construction* (*NDS*).[6] The *NDS*, first published in 1944, establishes requirements for structural wood construction. Most of these requirements focus on the establishing of minimum material design parameters and on the verification testing performed by the material producer. The *NDS* defines minimum fastener spacing and allowable loads for connections. Fastener spacing is typically measured periodically for compliance with *NDS* and the approved contract document requirements. At the same time, the inspector will look for lumber material markings that verify the producer has performed and accepted the *NDS*-specified verification testing.

10.1.3 Masonry

Building Code Requirements and Specification for Masonry Structures[7] provides the quality control requirements for masonry construction. The requirements are divided into two sets, Level B for risk categories I–III, and Level C for category IV. The main difference is that Level C requires more continuous inspection.

Masonry inspection requirements include:

- grout slump verification;
- masonry and grout compressive strength testing;
- proportions of site-prepared mortar;
- mortar joint construction;
- location and sizes of reinforcing;
- grout spaces are clear;
- size and location of structural elements;
- protection during cold or hot weather;
- placement of grout.

Trained personnel perform the inspections and prepare reports for the building official, architect, and structural engineer. Any defective work is removed and replaced with satisfactory construction.

10.1.4 Steel Fasteners

The specification of the Research Council on Structural Connections, *Specification for Structural Joints Using High-Strength Bolts*, includes the required testing of pre-tensioned and slip-critical fastener assemblies: nuts, bolts, and washers. Specialized axial tension testing machines measure the fastener clamping force placed on the machine by fully tightened assemblies. The specification establishes how many fasteners will be tested from each lot and the minimum tensile load they must resist. Fasteners installed in simple, bearing-type connections are installed in what is called the "snug-tight" condition. This means that the nut is tightened until all the clamped surfaces are in contact. Although not as stringent as the pre-tensioned or slip-critical fasteners, visual examination of these fasteners for proper identification, cleanliness, and so on, ensures the fasteners in the connection can perform as designed.

10.2 NONDESTRUCTIVE EXAMINATIONS OF WELDS

Welds are part of a family of production activities classically referred to as special processes. Other special processes include NDE, heat-treating, and coating systems. Because the end product or results of a special process cannot be actually tested without the workpiece being destroyed, the procedures for performing special processes must be proven by testing. Tension tests of sample welds can measure yield and ultimate tensile strength. Bend test specimens can be visually examined for ductility after being bent into a U-shape with code-specified bending fixtures. NDE technicians calibrate their NDE equipment against known standards to perform specified process checks that ensure the equipment will provide the resolution necessary to locate **discontinuities** and permit their evaluation. Special-process examination personnel are generally qualified by being tested to national standards that prove they can correctly implement the procedures.

The AWS establishes the methods and quality requirements for structural welding. AWS D1.1, *Structural Welding Code—Structural Steel*,[8] governs the vast majority of structural welding applications. AWS publishes other structural welding codes, including sheet metal, reinforcing steel, bridges, and so on. Each code has its own requirements and acceptance criteria appropriate to the welded end product being produced. All of these welding codes expect all welds to be completely visually examined, 100 percent; cracks, allowance for undersized weld, allowance for porosity, and weld profile requirements are all evaluated. All other NDE determined to be required to assure weld quality must be specified by the responsible engineer in the contract documents.

The American Society for Nondestructive Testing (ASNT), Recommended Practice SNT-TC-1A, *Personnel Qualification and Certification in Nondestructive Testing*,[9] establishes the recommended training and experience requirements for NDE technicians. SNT-TC-1A has long been the governing resource for qualifying NDE technicians. Level II or III NDE technicians are responsible for determining whether or not the indications they observe are rejectable. Acceptance or rejection is determined after the indications have been compared with the acceptance criteria specified in the governing standard or specification.

Breaks in the weld's expected surface condition, a surface otherwise continuous to the examiner's trained eye, are called discontinuities. The magnitude of the discontinuity and its frequency are compared with the product quality level and governing acceptance criteria so that it can be determined whether they are acceptable.

The majority of welding examinations are performed visually by simple observation of the weld's surface in the as-welded condition. Visual examination (VT) can be augmented by **liquid penetrant** testing (PT) or **magnetic particle** (MT) examinations. VT, PT, and MT are all "surface" examination methods. These additional examinations will highlight fine and small discontinuities that are easily overlooked when one uses only a flashlight for VT. PT and MT examination methods involve the application of additional dyes or colored particles to the examination surface. These collect around any discontinuities open to the surface and quickly capture the examiner's eye.

Volumetric examinations evaluate the weld through its entire volume. **Radiographic testing** (RT) and **ultrasonic testing** (UT) are the commonest NDE methods used in structural steel applications. RT uses gamma radiation, and UT uses sound that penetrates through the material's surface to locate discontinuities that can be scattered internally in the weld and base metal. Innovations in equipment now permit NDE indications to be digitized for later evaluation and review.

Cracks are not permitted in welds. They can **propagate** when they are driven by service loads or residual stresses. Similarly, lack of fusion between the weld and base metal or between weld passes is not tolerated and can also propagate under load. There is some tolerance for **porosity** and nonmetallic inclusions in the weld. These discontinuities are not prone to propagation. But, they do reduce the overall amount of metal in the joint, thus reducing the resulting strength of the weld. Porosity is generally limited to between 1 and 3 percent of the overall weld volume, depending upon the quality level required. Undersized welds are evaluated because of their effect on weld strength. Weld undercut and excessive weld reinforcement are welder workmanship issues. These discontinuities create stress risers at the outside edges of the welds that can reduce the ductility of the connection. There are other, subtler workmanship issues that are examined visually and are indicative of poorly controlled welding processes. For these reasons, all other NDE methods are performed after visual examination of the weld and surrounding base metal.

10.2.1 Liquid Penetrant Testing

PT is very portable and has broad application. The techniques are relatively simple and permit easy training. This NDE method can reveal discontinuities in nonferrous metals. It is also used on steel and other ferrous metals, when access by other NDE equipment is limited. Discontinuities must be open to the surface. When the penetrant is applied, it floods any cracks, porosity, incomplete weld fusion, or other surface discontinuities. Later, when the surface is cleaned of excess penetrant and the developer is applied, penetrant will be drawn out of any discontinuities into the developer by capillary attraction. The five process steps, shown in Figure 10.1, are as follows:

1. Clean the examination area.
2. Apply penetrant.
3. Remove excess penetrant.
4. Apply the developer.
5. Examine the area for penetrant indications.

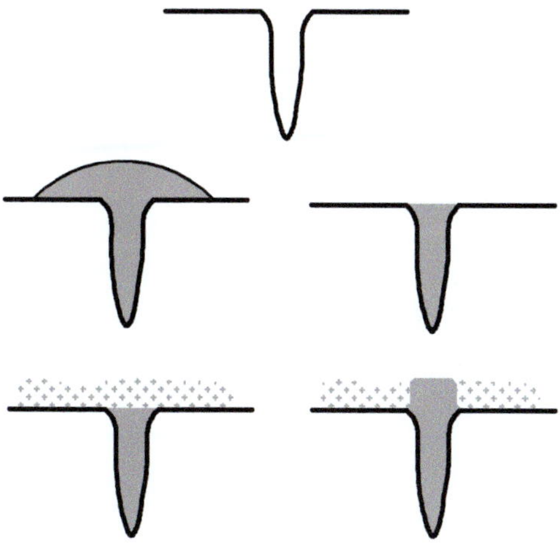

Figure 10.1 Liquid Penetrant Examination Sequence

There are two families of PT: fluorescent and visible-dye. Although the fluorescent method is the more sensitive of the two families, additional equipment is required, and any white light shining on the examination area must be kept below 2 foot-candles during evaluation. Fluorescent PT is commonly used in aerospace applications or on machined parts.

Visible-dye PT is more commonly used to examine nonferrous welds and the base metal around them. The penetrants are available in aerosol-spray and liquid-product forms, as are the specially formulated developers and cleaners. The method is inexpensive, very portable, and easy to use. PT is routinely used in both shop and field. Figures 10.2–10.6 illustrate the steps taken to examine welds in aluminum panels with the visible-dye PT examination method.

10.2.2 Magnetic Particle Testing

Magnetic particle examination offers a convenient and inexpensive means of examining ferromagnetic metals such as steels. The metal must be able to be magnetized. Thus, metals such as 300-series stainless steels and aluminum cannot be examined with this process. Several techniques can be used to magnetize the metal: bench units, high-amperage portable units,

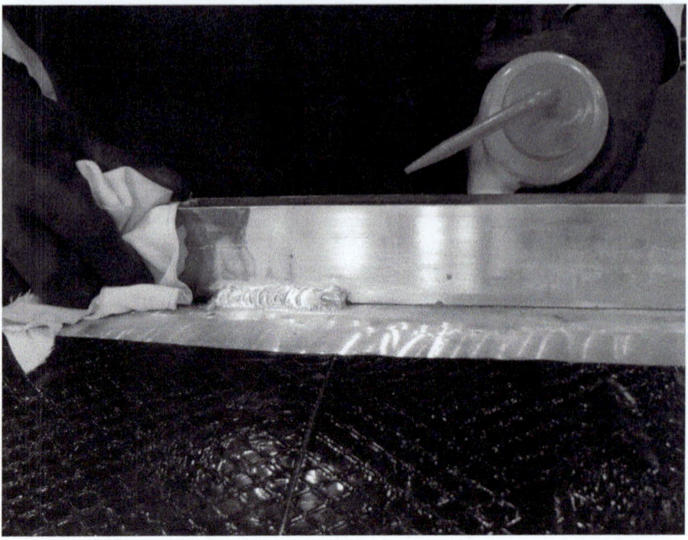

Figure 10.2 PT, Cleaning the Exam Area

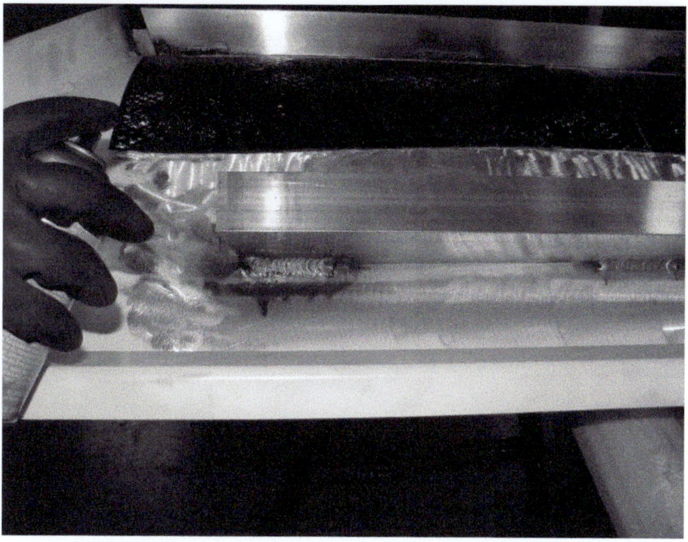

Figure 10.3 PT, Penetrant Application

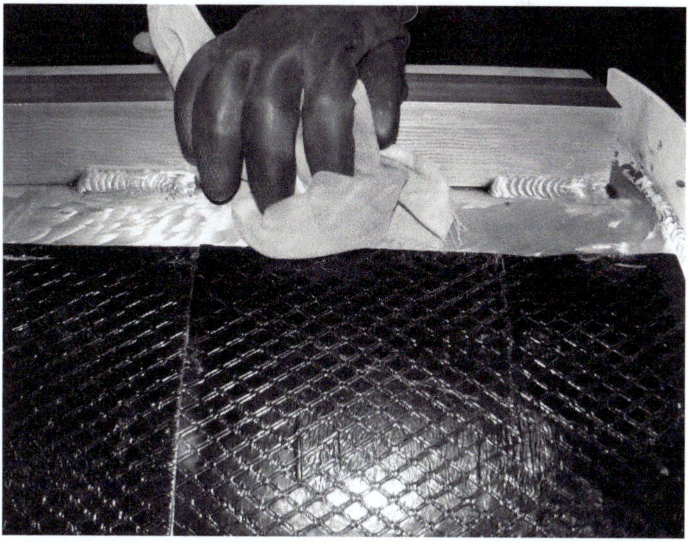

Figure 10.4 PT, Excess Penetrant Removal

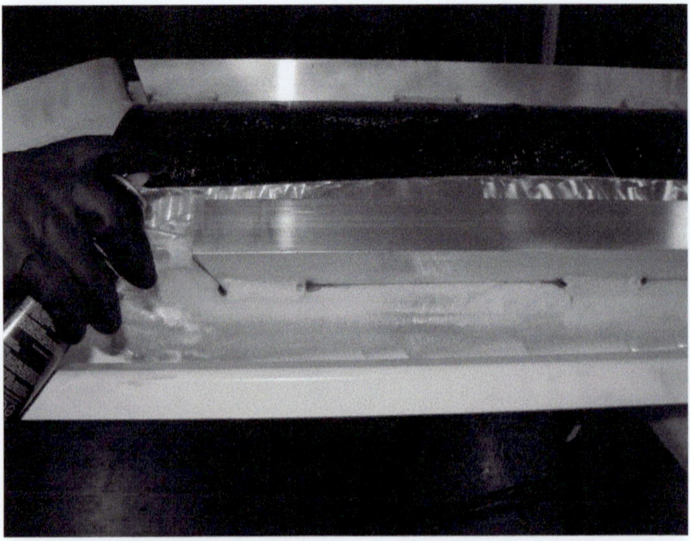

Figure 10.5 PT, Developer Application

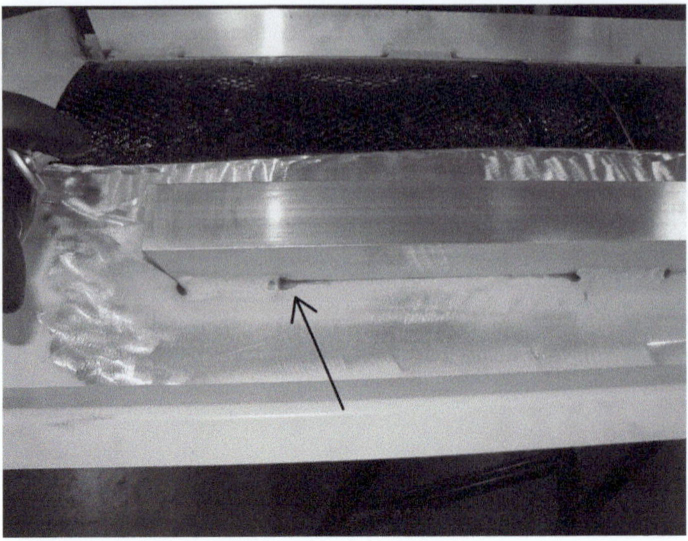

Figure 10.6 PT, Examination Results Exhibiting Porosity

prods, and yokes. The yoke method is, by far, most commonly used to examine structural steels.

Electromagnetic yokes require only a source of electricity to work, making them highly portable and efficient. Large areas can be examined, making the equipment suitable for examination of base-metal flaws, welds, joint preparations, and repair excavations. The process is, however, limited to surface discontinuities or, if direct current is used, discontinuities within ⅛ inch (3 mm) of the examination surface.

An electric current, pushed along a circuit by voltage, generates a field of magnetic flux around it. Faraday's law of electromagnetic induction predicts that increases in the current flow in the circuit will result in similar increases in the magnetic field surrounding it. Ampère's right-hand rule identifies the magnetic field running perpendicular to the direction of current flow. Magnetic fields used in structural steel applications are derived from current flowing along a straight line or flowing around a coil. The direction of the magnetic field will be either circular or longitudinal, respectively.

Bar magnets are permanently magnetized to create a longitudinal magnetic field. The molecules are aligned along an axis, creating positive and negative poles at the ends of the bar. Figure 10.7 illustrates how the magnetic field becomes distorted when the field leaves the bar and enters the surrounding air to continue its circuit to the opposite pole.

Ferromagnetic particles applied over the bar will collect at both its ends, at the metal-to-air field distortions. Breaking the bar magnet into two pieces will create new poles. The new ends will also collect magnetic particles. Finally, if a notch is cut into a bar magnet, the discontinuity will force a distortion in the magnetic field, and magnetic particles will collect along its edges.

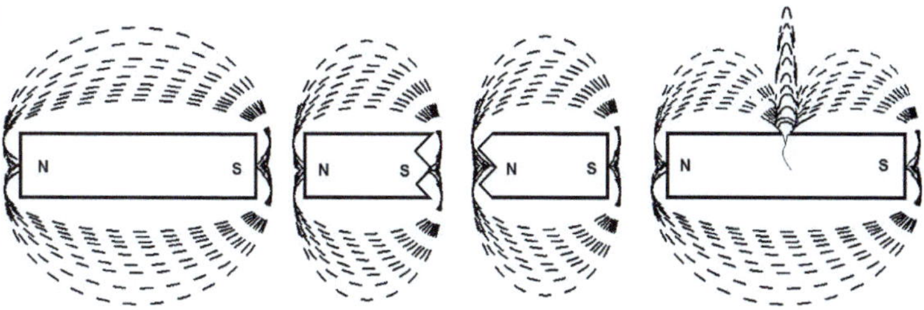

Figure 10.7 Distortions in a Magnetic Field

When weld joints are magnetized, surface discontinuities also create local poles. If the magnetic particles are colored, or made fluorescent, this enhances identification of the observed discontinuities for the technician. The magnetic distortions are readily observed when magnetic particles are applied and collect at the discontinuities.

MT technicians have various methods available to magnetize components in several directions, to assure 100 percent examination coverage. Cracks and other discontinuities are best observed when they are aligned perpendicular to the magnetic field. The strength of the magnetic field drops off as the discontinuity's alignment varies from true perpendicular. Beyond 45°, the discontinuity is not being magnetized with sufficient strength to hold magnetic particles as required by the governing codes. For this reason, parts must be magnetized in at least two perpendicular directions to assure 100 percent coverage of the examination area.

For electromagnetic yokes, the examination area is limited to the clear spacing between the legs of the yoke and approximately 1½ in (38 mm) on either side. Multiple magnetizing shots are taken. Powders of iron particles, applied in a light cloud over the area, will be magnetically attracted to breaks in the surface continuity and the field distortions surrounding them. Figure 10.8 shows the yoke method on a large fillet weld. Figure 10.9 shows a crack running along the weld and the tubular base metal. This crack was identified using the yoke position shown in Figure 10.8.

10.2.3 Radiographic Testing

RT is one of the common volumetric examination methods. It is a photographic method that uses radioactive sources to provide the energy that is recorded on photographic film or on a variety of digital imaging devices. Film is most commonly used, because of its convenience and because a physical record of the indications remains after the examination is finished. RT technicians are able to examine the entire volume of material between the radioactive source and the imaging device. They evaluate the images, looking for changes in film densities, which reflect discontinuities from the surrounding material—lighter and darker areas in the exposure.

Cracks create voids in metal as it separates and the cracks run. Voids are not as dense as the surrounding metal. They permit more radiation to pass through to the film. The RT technician observes them to be darker

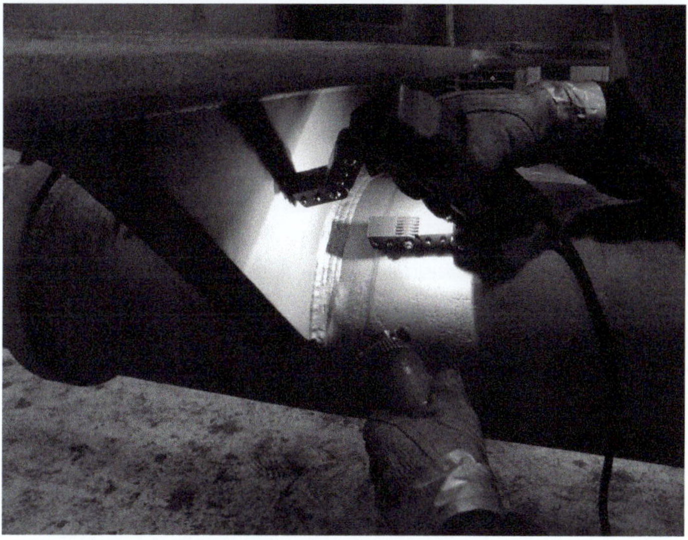

Figure 10.8 Magnetic Particle Testing—Yoke Method

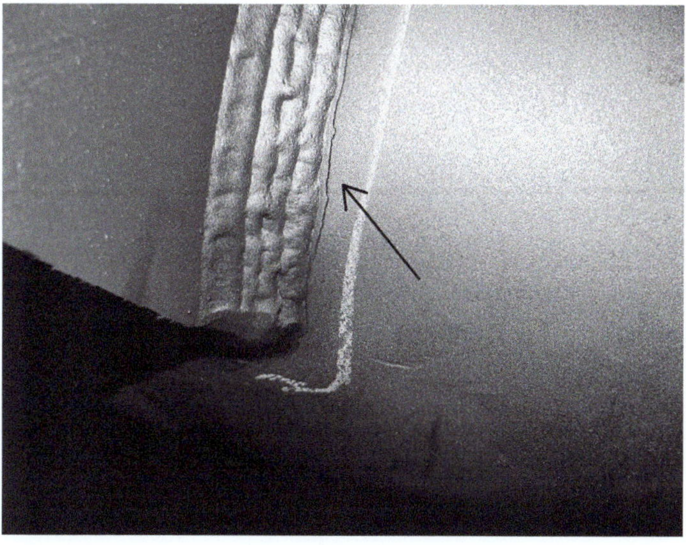

Figure 10.9 Magnetic Test Indications Showing a Crack

regions in the radiograph. Similarly, nonmetallic inclusions and pockets of gas porosity are less dense, and the penetrating radiation passes through the discontinuity easily. Figure 10.10 shows the dark bubbles of gas porosity in the lighter and denser weld metal. This figure also shows the image quality indicator RT technicians use to verify the resolution and acceptability of their radiography technique. The faint white lines, beneath the **ASTM** at the top of the figure, verify the technique is capable of identifying discontinuities requiring evaluation and judgment. Denser areas, such as the steel reinforcing bars cast in concrete or tungsten inclusions in welds, will appear as lighter areas.

This NDE method has a long history of successful application. Concrete decks are radiographed to assure post-tensioning cables will not be cut during core-drilling for drains and other deck penetrations. Steel welds in pressure vessels are routinely X-rayed so that the entire volume of weld metal can be examined. RT offers reasonable assurance that the weld and surrounding weld zone are sufficiently free of mechanical discontinuities to perform as designed.

X-ray is only one source of radiation used to make exposures. Iridium-192 and cobalt-60 gamma-ray sources are common radioactive isotopes used in industrial radiography. Cobalt-60 is a highly penetrating radiation source. It is used to radiograph thick sections of concrete and steels. Iridium-192 is most often used to radiograph structural welds. Structural welds, however, are seldom examined by radiography. The radiation these

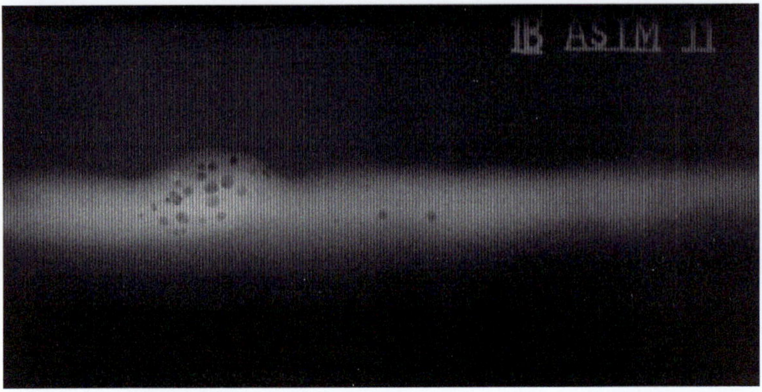

Figure 10.10 Radiograph Showing Weld Porosity

sources emit is dangerous to humans. Other work inside the safety perimeter must stop while RT is being conducted. For this reason, most structural welds are examined with ultrasonic testing.

10.2.4 Ultrasonic Testing

Finally, UT is commonly used for volumetric examination of welds and materials in industry. UT is often the preferred method for examining complete joint penetration, full-strength welds. There are no unusual safety concerns using the equipment, and other work in the area is not impacted. Two kinds of UT are performed today, conventional UT and its recently developed variation, phased-array UT (PAUT).

Sonic impulses are introduced into the examination area when this examination is performed. The sound waves they generate will reflect back to their point of entry when they hit internal discontinuities. UT equipment incorporates an imaging device and a transducer, or probe. The imaging device furnishes electrical signals that excite piezoelectric elements installed in the probe. The element contracts when electrically charged and returns to its original shape when turned off. The "snap" to its original position provides the mechanical energy needed to generate the sound wave into the examination area. Returning sound waves, reflected back from free material surfaces and internal discontinuities, cause the piezoelectric element to contract, which produces voltage that registers on the imaging device. The imaging device includes filtering and signal adjustment controls, as well as the software needed to interpret the reflected signals.

Conventional UT equipment uses single-angle probes to transmit sound into the examination area. This technique has been proven to effectively identify discontinuities in base metals and welds that might violate code requirements. Figure 10.11 shows a technician verifying the UT equipment set-up quality by scanning a calibration block with a known flaw machined into it.

Conventional UT is capable of producing three imaging modes, but the imaging devices are relatively simple to use. UT has enjoyed broad use over the years. The method is routinely used to measure pitting in pressure vessels and pipes, scan for lamination or stringers in base metals, and examine welds in structural connections. Figure 10.12 combines the probe with the imaging device being used to examine a weld.

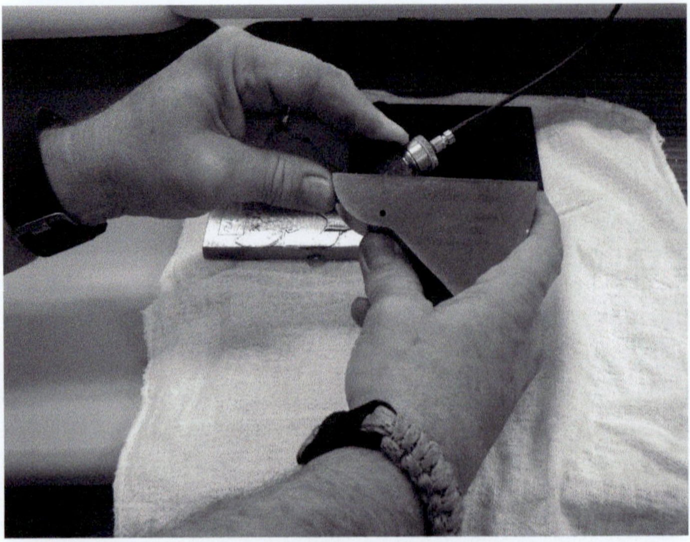

Figure 10.11 Conventional UT, Probe, and Calibration Standard

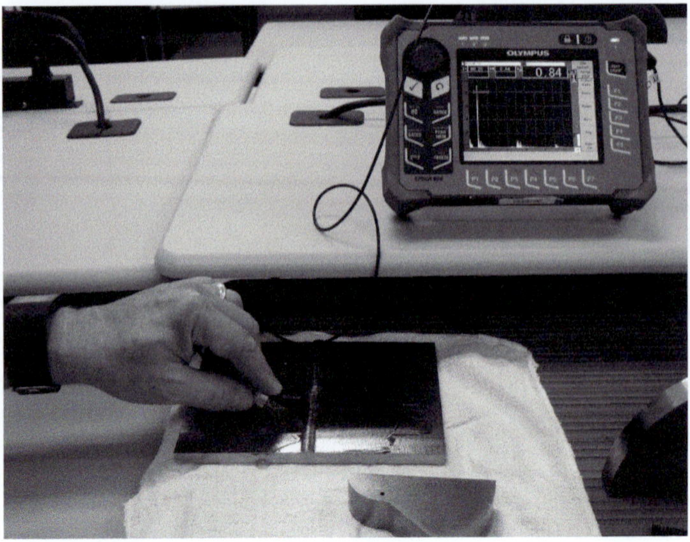

Figure 10.12 Conventional UT, Probe, and Imaging Device

PAUT equipment combines multiple piezoelectric elements into one probe. Where UT uses only single- or dual-element probes, PAUT may use 16, 32, or up to 256 individual elements in one probe. PAUT imaging devices include sophisticated computer software that finely tunes the multi-element probe. Elements can be programmed to fire at different angles, in different sequences, and at different focal points, to improve focusing of the resulting sound wave and to "steer" them to focus on specific points of interest in the examination area. Proprietary software resolves the reflected sound waves into five specialized visual imaging modes, unique to PAUT, for evaluation by the UT technician. A calibration artifact is shown in Figure 10.13, and the reflections of the sound waves are shown on the imaging device screen. PAUT is a relatively new NDE method and new PAUT examination techniques are continually being discovered.

UT technicians evaluate the variations in the amplitude of the high-frequency sound waves they observe on the imaging device, reflected back to the probe element. Reflections will occur at the near and far surfaces of materials, at fixed ends, at changes in cross sections, and at internal material discontinuities. High-amplitude spikes indicate sharply defined discontinuities, such as cracks or laminations. Low-amplitude spikes indicate poor reflectors, such as porosity. By adjusting the grids on the imaging device to clearly identify the amplitude peaks caused by the free surfaces, the technician can accurately calculate the location and depth of any internal discontinuities reflecting sound waves.

Technicians calculate their sound paths based upon the incident angle of the sound entering the material. Conventional UT probes are typically manufactured in fixed angles: 45°, 60°, and 70°. Projecting the sound into material at an angle as a shear wave enables some of the sound waves to reflect back into the material, much like a pool ball bouncing off the bumper surrounding a pool table. This second "leg" is reflected back into the examination area from the far surface, opposite the sound entry point. In the case of welds and weld zones, the technician can plot the sound path needed to examine the entire volume of the weld zone with the selected UT probes. Combining straight-beam, through-section UT examinations with the shear-wave examinations assures 100 percent coverage of the examination area.

10.3 CONCLUSION

All governing codes and standards mandate minimum inspection points, examinations, and tests that must be performed by fabricators and

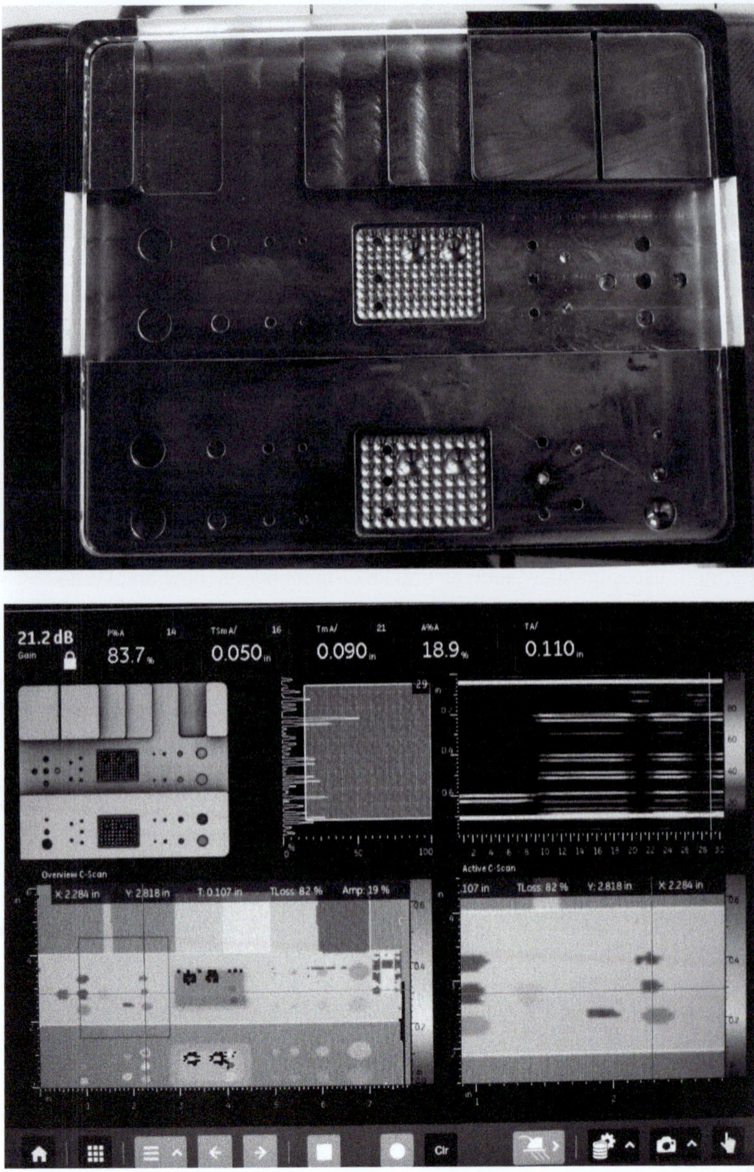

Figure 10.13 PAUT Calibration Artifact and Screen Image

installers. Architects and responsible engineers may remove or revise any imposed requirement. They are aware, however, that their decisions come at the price of increased risk and potential legal liability. With the cost of noncompliance so high, experienced designers approach any deviation from national standards and local regulations with caution. Today's work environment expects engineers and contractors alike to know the work has been done correctly, not hope.

NOTES

1. P. Crosby. *Quality Is Free: The Art of Making Quality Certain* (New York: McGraw-Hill, 1979).
2. ICC. *International Building Code* (Washington, D.C.: International Code Council, 2012).
3. Clark County Building Department. *Preparation of a Quality Systems Manual for Structural Steel Fabricators*, TG-1–06a (Clark County, NV: Clark County Building Department, 2006).
4. ACI. *Building Code Requirements for Structural Concrete*, ACI 318 (Farmington Hills, MI: American Institute of Concrete Construction, 2014).
5. ACI. *Specification for Tolerances for Concrete Construction and Materials*, ACI 117 (Farmington Hills, MI: American Institute of Concrete Construction, 2010).
6. AWC. *ASD/LFRD Manual, National Design Specifications for Wood Construction* (Leesburg, VA: American Wood Council, 2012).
7. TMS. Building Code Requirements and Specification for Masonry Structures, TMS 402–13 (Longmont, CO: The Masonry Society, 2013).
8. AWS. *Structural Welding Code—Steel*, D1.1 (Miami, FL: American Welding Society, 2015).
9. ASNT. *Personnel Qualification and Certification in Nondestructive Testing*, SNT-TC-1A (Columbus, OH: American Society for Nondestructive Testing, 2016).

List of Units

Appendix 1

Table A1.1 List of Units - Metric

Unit	*Definition*	*Typical Use*
°	Degrees	Angle
deg	Degrees	Angle
g	Gram	Mass
hr	Hour	Time
km/h	Kilometers per hour	Speed
kN	Newton	Force
kN	kiloNewton	Force
kN/m	kiloNewton per meter	Distributed linear force
kN/m^2	kiloNewton per square meter (aka kPa)	Distributed area force, pressure
kN/m^3	kiloNewton per cubic foot	Density
kN-m	kiloNewton-meter	Moment, torque
m	Meters	Length
m^2	Square meters	Area
m^3	Cubic meters	Volume
min	Minute	Time
mm	Millimeters	Length
mm^2	Square millimeters	Area
mm^3	Cubic millimeters	Volume
mm^4	Millimeters to the fourth power	Moment of inertia
MN/m^2	kiloNewton per square inch (aka GPa)	Distributed area force, pressure
N	Newton	Force
N/m	Newtons per meter	Distributed linear force
N/m^2	Newtons per square meter (aka Pa)	Distributed area force, pressure
N/m^3	Newtons per cubic meter	Density
N/mm^2	Newtons per square millimeter (aka MPa)	Distributed area force, pressure
Pa	Newton per square meter (N/m^2)	Distributed area force, pressure
rad	Radian	Angle

Table A1.1 List of Units - Imperial

Unit	*Definition*	*Typical Use*
°	Degrees	Angle
deg	Degrees	Angle
ft	feet	length
ft^2	square feet	area
ft^3	cubic feet	volume
hr	hour	time
in	inches	length
in^2	square inch	area
in^3	cubic inch	volume
in^4	inches to the fourth power	moment of inertia
k	kip (1000 pounds)	force
k/ft	kips per foot (aka klf)	distributed linear force
k/ft^2	kips per square foot (aka ksf)	distributed area force, pressure
k/ft^3	kips per cubic foot (aka kcf)	density
k/in^2	kips per square inch (aka ksi)	distributed area force, pressure
k-ft	kip-feet	moment, torque
lb, lb_f	pound	force
lb/ft	pounds per foot (aka plf)	distributed linear force
lb/ft^2	pounds per square foot (aka psf)	distributed area force, pressure
lb/ft^3	pounds per cubic foot (aka pcf)	density
lb/in^2	pounds per square inch (aka psi)	distributed area force, pressure
lb-ft	pound-feet	moment, torque
rad	radian	angle
yd^3	cubic yard	volume

List of Symbols

Appendix 2

Table A2.1 List of Symbols

Symbol	*Definition*	*Units*	
		Imperial	*Metric*
c'	Cohesion	lb/ft^2	kN/m^2
c	Half crack length	in	mm
δ	deflection	in	mm
f	Induced or forcing frequency	rpm	Hz
f	Far field stress	lb/in^2	MN/m^2, MPa
f_n	Natural frequency	rpm	Hz
g	Acceleration of gravity	in/s^2	m/s^2
γ	Unit weight (density)	lb/ft^3	kN/m^3
γ	Surface tension	lb/in	N/m
k	Stiffness	lb/in	N/m
LL	Liquid limit	Unitless	
m	Mass	lbm	kg
PI	Plasticity index	Unitless	
PL	Plastic limit	Unitless	
q	Effective stress	lb/ft^2	kN/m^2
q_a	Allowable bearing capacity	lb/ft^2	kN/m^2
$q_{a(net)}$	Net allowable bearing capacity	lb/ft^2	kN/m^2
$q_{net(u)}$	Net ultimate bearing capacity	lb/ft^2	kN/m^2
q_u	Ultimate bearing capacity	lb/ft^2	kN/m^2
q_q	Overburden	lb/ft^2	kN/m^2
R_d	Dynamic amplification factor	Unitless	
u_o	Dynamic displacement	in	mm
u_{st}	Static displacement	in	mm
ξ	Damping ratio	Unitless	

Imperial and Metric Conversion Tables

Appendix 3

Table A3.1 Imperial and Metric Conversion Tables

Imperial to SI					
Multiply		*By*		*To get*	
	ft		0.305		m
	ft^2		0.093		m^2
	ft^3		0.028		m^3
	in		25.4		mm
	in^2		645.2		mm^2
	in^3		16,387		mm^3
	in^4		41,6231		mm^4
	k		4.448		kN
	k/ft		14.59		kN/m
	k/ft^2		47.88		kN/m^2
	k/ft^3		157.1		kN/m^3
	k/in^2 (ksi)		6.895		MN/m^2 (MPa)
	k-ft		1.356		kN-m
	lb, lb_f		4.448		N
	lb/ft		14.59		N/m
	lb/ft^2 (psf)		47.88		N/m^2 (Pa)
	lb/ft^3		0.157		kN/m^3
	lb/in^2		6,894.8		N/m^2
	lb-ft		1.355		N-m
	lb_m		0.454		kg
	mph		1.609		kmh

Table A3.1 Continued

SI to Imperial					
Multiply	m	*By*	3.279	*To get*	ft
	m^2		10.75		ft^2
	m^3		35.25		ft^3
	mm		0.039		in
	mm^2		0.0016		in^2
	mm^3		6.1E-05		in^3
	mm^4		2.4E-06		in^4
	kN		0.225		k
	kN/m		0.069		k/ft
	kN/m^2		0.021		k/ft^2
	kN/m^3		0.0064		k/ft^3
	MN/m^2 (MPa)		0.145		k/in^2 (ksi)
	kN-m		0.738		k-ft
	N		0.225		lb, lb_f
	N/m		0.069		lb/ft
	N/m^2 (Pa)		0.021		lb/ft^2 (psf)
	kN/m^3		6.37		lb/ft^3
	N/m^2		1.45E-04		lb/in^2
	N-m		0.738		lb-ft
	kg		2.205		lb_m
	kmh		0.621		mph

Glossary

AAMA	American Architectural Manufacturers Association
abatement	removal or containment of hazardous materials
acceleration	increase in the speed of something
acceleration time histories	record of acceleration over time from a specific earthquake event
accelerometer	device that senses acceleration
acceptance requirements	criteria determining the acceptance or rejection of test results
access control	control of people, vehicles, and materials through the implementation of security measures for a protected area
acidification	ongoing decrease in the pH level of the Earth's oceans
adaptability	the ability to change function or purpose from original intended design
aerodynamic	way air flows around a structure
aesthetics	sense of beauty usually related to visual appeal
aggregate	the stones and sand (course and fine aggregate, respectively) used as a filler in concrete, asphalt, etc.
air-blast loading	pressure applied to an object located within the path of a blast wave propagation
alkali–silica reactivity	reaction between cement and aggregate, causing cracking

allowable bearing capacity	the gross ultimate bearing capacity divided by a factor of safety
allowable net bearing capacity	the net ultimate bearing capacity divided by a factor of safety
anode	negative element in a corrosion cell, where corrosion occurs
ASTM	American Society for Testing and Materials
Atterberg limits	the collective term for the liquid, plastic, and shrinkage limits of a soil sample
authentic sustainable design	design that goes beyond energy conservation to look holistically at cultural, societal, and ecological issues and beauty
axial	action along length (long axis) of member
backbone curve	curve showing normalized yield behavior of a structure
balanced design	controlled failure of a system with an established hierarchy of component failures
balsa wood	native wood of South American rain forests, known for its light-weight strength
base isolators	rubber and steel pads that shift a building's period, resulting in lower seismic accelerations
basis of design	narrative that is provided to all design teams, with specific instructions to be implemented into the design
beam	horizontal member resisting forces through bending
biodiversity	the variety of life in the world or in a particular habitat
biomimicry	the design and production of materials, structures, products, and systems modeled on nature's patterns and strategies
blast	synonym for explosion
blast mitigation measures	measures that remove or reduce exposure or vulnerability to blast

blast-resistant design	design of the structural and nonstructural elements of a building to resist the effects of a blast event
braced frame	structural frame whose lateral resistance comes from diagonal braces
brittle	material with low deformation characteristics
buckling	excess deformation or collapse at loads below the material strength
building envelope	separation between the interior and the exterior environment of a building
capacity	ability to carry load, related to strength of a member
carbon emissions	carbon dioxide produced by planes, cars, factories
carbon fiber	high-strength filament fiber
carbon footprint	total amount of carbon dioxide produced directly or indirectly by a particular person, group, event, or product
cathode	positive element in a corrosion cell, protected area
cementitious	being made up of cement or pozzolanic material
chloride	salts that accelerate reinforcing steel deterioration
cladding	the separately applied exterior finish of a framed building
CO_2	carbon dioxide
code	compilation of rules governing the design of buildings and other structures
cohesion	an electrostatic force that draws clay soil particles together
collaborative design	design including all the stakeholders and professionals, where information sharing is unrestricted
collapse prevention	performance level intended to leave the building barely stable after an event

collapsible soils	soils that display high strength while dry, but can lose strength rapidly with wetting
column	vertical member that primarily carries axial compression load, supports floors and roofs
component	single structural member or element
compression	act of pushing together, shortening
condensation	water droplets that collect on a cold surface in contact with humid air
cone penetrometer test	standard test method to measure in situ soil compaction without taking samples
connection	region that joins two or more members (elements)
constraint	restraint of material against movement
continuous operation	performance level intended to ensure the building is functional immediately after an event
corrosion	chemical reaction resulting in the gradual destruction of metals
couple, or force couple	parallel and equal, but opposite, forces, separated by a distance
creep	slow, permanent material deformation under sustained load
criterion	standard upon which decisions are made
cross laminated timber	wood product formed by gluing or dovetailing wood boards, in alternating directions
damping	decrease in amplitude of an oscillation as energy is drained from the system
dead load	weight of permanent materials
deflection, δ	movement of a member under load or settlement of a support
deforestation	removing trees to transform a forest into cleared land without consideration of ecological factors
demand	internal force due to applied loads

design basis threat (DBT)	explosive type and charge size for which the building is intended to provide a specified level of protection
destructive test	test that results in destruction of specimen
deterioration	process of becoming worse over time
disproportionate collapse	collapse characterized by a disproportionate relationship between the event and structural damage
discontinuity	interruption in material, such as a crack or porosity
drift	lateral displacement between adjacent floor levels in a structure
dual-pane unit	glazed unit made up of two panes of glass
ductile, ductility	measure of the capability of an element to undergo significant yielding without a reduction in strength
durability	ability to resist deterioration
dynamic	object in motion or vibrating
dynamic amplification factor	factor multiplied by static deflection to produce dynamic deflection
eccentricity	offset of force from centerline of a member, or centroid of a fastener group
ecological	concerning the relationship of living organisms of all scales to one another and their physical surroundings
ecosystem	all the physical and biological components of an environment working as an interactive system
elasticity	ability to return to original shape after being loaded
electrolyte	liquid capable of conducting electrical current
electron flow	flow of electricity
element	single structural member or part

embed a structural support that has been fully incorporated into another material such as concrete or stone

embodied energy energy consumed by all of the processes associated with production, from resource extraction through product delivery

envelope the physical separation between exterior and interior environments

environment temperature, weather, and chemical conditions in which a structure is located

environmental footprint measure of human impact on the Earth's ecosystems

event noteworthy happening

expanded metal lathe a thin metal sheet that has been slit and expanded to produce diamond-shaped holes

expansive soils plastic clay soils that display excessive shrinkage and swelling with the subtraction and addition of moisture

explosion rapid chemical reaction that produces noise, heat, and violent expansion of gas

façade any side of a building

fascia any relatively broad, flat, horizontal surface, as the outer edge of a cornice, a stringcourse, etc.

fast fracture crack moving at the speed of sound in a material

fatigue time-dependent crack growth due to cyclic stresses

fatigue threshold theoretical stress limit below which fatigue does not occur, though not proven by experience

fiber-reinforced polymers (FRP) composite made by combining a plastic polymer resin with carbon, glass, or aramid fibers

finite element analysis (FEA) analytical tool that uses stiffness matrices to calculate forces and deflections

fire effects	thermal and structural response due to fire exposure and subsequent cooling
fire exposure	the extent to which materials, products, or assemblies are subjected to the conditions created by fire
fire resistance	the time duration a mock-up assembly is able to withstand furnace heating
fire sprinkler system	fire suppression system that automatically discharges water over a specified area
fixed	boundary condition that does not permit translation or rotation
flexure, flexural	another word for bending behavior
footing	foundation system bearing on soil near the ground surface
force	effect exerted on a body
forcing frequency	the rate at which a dynamic force is applied
fracture	separation of two materials that were once together
fracture mechanics	study of crack propagation in materials
frame	system of beams, columns, and braces, designed to resist vertical and lateral loads
frequency	number of recurrences of a repeating event in a given time, cycles per unit time
friction angle	the angle of internal friction of a soil sample that provides one component of shear strength
fundamental period	first period of vibration accounting for the primary structural response
galvanic series	chart relating the corrosion potential between different metals
general shear failure	sudden shear failure that occurs when the soil reaches its ultimate bearing capacity; occurs in densely compacted soils
girder	beam that supports other beams

global warming potential	comparison of the amount of heat trapped by a certain mass of the gas in question with the heat trapped by the same mass of carbon dioxide
grain	individual crystal in metal
grain size distribution	the relative proportions of each particle size in a soil sample
gravity load	weight of an object or structure, directed towards the center of the Earth
green plateau	where buildings are mitigating their impact but not playing a part in regenerating the ecosystem
greenbelt land	policy and land use designation to retain areas of undeveloped, wild, or agricultural land
gross ultimate bearing capacity	the maximum total loading on a soil layer before shear failure
harmonic	repeating load or wave function
heat transfer	the exchange of thermal energy due to a temperature difference
high-cycle fatigue	fatigue occurring over millions of cycles, below the yield stress
historic fabric	original construction of historical importance
holistic	dealing with the whole of a thing as intimately interconnected
immediate occupancy	performance level intended to ensure structure is occupiable immediately after an event
induced frequency	see forcing frequency
inelasticity	behavior that goes past yield, resulting in permanent deformation
integral theory	framework used to organize the multilevel needs in sustainable design
integrated design process	collaborative design process where all design disciplines are included at the beginning of the process

lamination	process of overlaying multiple layers of composite material for strength, stability, or appearance
lateral load	load applied in the horizontal direction, perpendicular to the pull of gravity
ledger	member supporting another material
life cycle analysis	assessment of the environmental impact of all stages of a product's life, from raw resource extraction to delivery to end of product life
life safety	performance level intended to protect human life after an event
light transmission	the ability of light to move through a surface or material
linear dynamic	analysis method based on dynamic forces and linear material behavior
linear static	analysis method based on static forces and linear material behavior
liquid limit	the moisture content that defines the difference between a plastic and semi-liquid soil state, defined by ASTM D4318
liquid penetrant	weld test using dyes to detect cracks
live load	load from occupants or moveable building contents
load	force applied to a structure
load path	route a load takes through a structure to reach the ground
local shear failure	gradual shear failure that occurs when soil reaches its ultimate bearing capacity
low-cycle fatigue	fatigue occurring in tens of thousands of cycles, often above the yield stress
low-e coating	a material that emits low levels of radiant thermal energy
magnetic particle	weld test using magnetic fields and iron particles to detect cracks

mass	measure of an object's resistance to acceleration, weight divided by acceleration of gravity
materiality	quality or character of material used
microclimate	climate of a small area that is different from the general climate
modal analysis	analysis method based on structure vibration modes and associated acceleration
mode	standing wave state of excitation for a given frequency
mode shape	vibration shape with a specific natural frequency
modular	composed of standardized units or sections for easy construction or flexible arrangement
module	distinct but interrelated unit that can be combined into a complex program
moisture content	see water content
moment, *M*	twisting force, product of force and the distance to a point of rotation
mortar	a mixture of lime or cement with sand and water
natural frequency	the rate at which a structure vibrates
net ultimate bearing capacity	the maximum structure weight a foundation can support; ultimate bearing capacity minus overburden
nonconductive	inert material designed to impede the flow of heat
nondestructive examination	test that does not destroy the specimen
nonlinear dynamic	analysis method based on dynamic forces and nonlinear material behavior
nonlinear static	analysis method based on static forces and nonlinear material behavior
nucleation	the act of beginning to form

occupancy	type of building use by people
optical fiber	flexible, transparent fiber made of glass or plastic, slightly thicker than a human hair
optimum moisture content	the moisture content at which maximum compaction can be achieved
orthogonal	intersections that are perpendicular or at right angles
overburden	the weight of a foundation plus any additional soil overlying the foundation
overpressure	pressure caused by a shock wave over and above normal atmospheric pressure
performance-based design	design method based on ultimate performance rather than prescriptive rules
performance levels	varying levels of survival after an event
pin	boundary condition that allows rotation but not translation
plane strain	condition where strain is zero in one direction, perpendicular to applied stress, associated with thicker material
plastic limit	the moisture content that defines the difference between a plastic and semi-solid soil state, defined by ASTM D4318
plasticity index	the difference between the liquid limit and the plastic limit
poorly graded soil	soil with a small range of particle sizes represented
porosity	the ratio of volume of voids to total volume of a soil sample, expressed as a percentage; small holes in what should be solid material
pressure	force per unit area
progressive collapse	chain-reaction failure of a building's structural system or elements as a result of initial localized damage
propagate	spread or grow

protection level	qualitative degree to which a building is expected to prevent or limit injury to and fatality of its occupants, and damage to and destruction of its contents, in the event of an explosion
protective design	design incorporating measures to improve security and protection of an asset
PTFE fabrics	synthetic materials coated and laminated to increase strength, durability, and environmental resistance
punching failure	shear failure that occurs owing to soil consolidation; occurs in loosely compacted soils
pushover analysis	see nonlinear static
quality	conformance to requirements, degree of excellence
racking	the distortion of a rectangular shape to a skewed parallelogram
radiography	weld test using X-rays to detect cracks
reaction	force resisting applied loads at end of member or bottom of structure
redundant	redundant system has spare capacity purposely created to accommodate disruption or surges in demand
reflected impulse	cumulative blast loading over time, calculated as the area beneath the reflected pressure–time curve
reflected pressure	increased blast pressure due to change in momentum of air molecules close to the surface of a structure
resilience	will and ability to anticipate, endure, adapt, and thrive within a disruptive and changing environment; capability to recover to full functionality after an extreme event
resonance	condition where external, oscillating force drives another system to oscillate at a greater amplitude at a preferential frequency
resources	stock of supply

retrofit	adapt to a new purpose, often correcting deficiencies
risk	function of the severity of a hazard and exposure to that hazard
rupture	complete separation of material
safety factor	factor taking into account material strength or load variability
safety objective	combination of seismic risk and performance level
saturated	soaked, impregnated, or imbued thoroughly
saturation	the ratio of volume of water to volume of voids in a soil sample, expressed as a percentage
scaled distance	ratio of the stand-off to the cube root of the explosive charge size as an equivalent mass of TNT
sealant	material used for sealing joints between materials so as to make them airtight or watertight
seismic design category	classification based on occupancy and earthquake severity
seismic hazard	level of seismicity, often expressed as potential ground-shaking intensity
seismic load	force accounting for the dynamic response of a structure or element due to an earthquake
seismic-force resisting system	portion of structure designed to resist earthquake effects
shear	equal, but opposite, forces or stresses, acting to separate or cleave a material, like scissors
shear wall	wall providing lateral resistance for structure
shearing layers	set of building components that evolve in different time scales
shrinkage limit	the moisture content that defines the difference between a semi-solid and solid soil state, defined by ASTM D427
sinusoidal	loading applied as a sine function

soil spring method	a foundation design analysis method that models various soil strata with varying springs
spandrel	a horizontal band of cladding that occurs at floor or roof levels of a building, usually opaque
special inspections	quality inspections required by the building code
stability	structure's resistance to excessive deformation or collapse at loads below the material strength, opposite of buckling
standard fire resistance design	the selection of fire-resistive assemblies to meet code requirements for structural fire resistance, based on their performance in standard furnace testing
standard penetration resistance	*N*-value as determined by the standard penetration test; used to quantify in situ soil compaction
standard penetration test	ASTM D1586; standard test method to measure the standard penetration resistance value
stand-off	physical distance between the surface of a building and the potential location for an explosion
stiffness	resistance to deformation when loaded
Stoke's law	calculation of terminal velocities of particles moving through a fluid medium
strain rate	rate of applied loading
strength	material or element resistance to load or stress
stress (f, t)	force per unit area
stress intensity	stress and geometry parameter related to a crack
structural analysis	determination of forces, moments, shears, torsion, reactions, and deformations due to applied loads
structural fire engineering	the explicit design of structural members and connections to satisfy performance objectives for structural design fires

structural integrity	ability of a structural member to perform a given function under load and environmental conditions
structural system	series of structural elements (beams, columns, slabs, walls, footings) working together to resist loads
sulphate attack	salts that attack the cement paste, leading to concrete deterioration
superstructure	the primary structural support system of a building
support	either the earth or another element that resists movement of the loaded structure or element
sustainability	ability to be self-supporting without depleting natural resources or ecological balance
sustainable design	philosophy of designing the built environment to reduce consumption of nonrenewable resources, minimize waste, and create healthy, productive environments, socially, economically, and ecologically
target displacement	desired displacement in analysis showing the structure is stable under the given seismic forces
tension	act of pulling apart, lengthening
thermal restraint	a condition in which thermal expansion or contraction of structural members is resisted by forces external to the members
threat	action or event that could result in a loss, or an indication that such an action or event might take place
tolerance	the permissible range of variation in a dimension of an object or objects
toughness	material resistance to cracking
transient	load lasting only for a short time
translucent	transmitting light but causing enough diffusion to prevent distinct images
triaxial stress	stress in three orthogonal directions

triple-pane unit	glazed unit made up of three panes of glass
ultimate bearing capacity	the loading per unit area of foundation at which a soil experiences shear failure
ultrasonic	weld test using ultrasonics (sound waves) to detect cracks
universal method	see soil spring method
urban context	the interrelating conditions that form the setting of a city or town
urban sprawl	uncontrolled expansion of urban development away from centralized areas into low-density land
value engineering	systematic and organized approach to provide necessary functions in a project at the lowest cost
vehicle security barrier	barrier providing a hard stop to encroachment and penetrative vehicle-borne attacks
vibration	periodic motion of an object forced from a position of equilibrium
viscous dampers	shock absorbers that dissipate energy
void ratio	the ratio of volume of voids to volume of soil particles, expressed as a percentage
vulnerability	exploitable weakness
water content	the ratio of the weight of water to the weight of dry soil in a soil sample
wear	damage to a solid surface due to relative motion between the surface and a contacting material
weldable	ability to be welded to a steel material
well-graded soil	soil with a wide range of particle sizes represented
yield, yielding	point at which a material has permanent deformation owing to applied loads, start of inelastic region of stress–strain curve

Bibliography

AAMA. *Recommended Dynamic Test Method for Determining the Seismic Drift Causing Glass Fallout from a Wall System*, AAMA 501.6–09 (Schaumburg, IL: American Architectural Manufacturers Association, 2009).

AAMA. *Recommended Static Test Method for Evaluating Curtain Wall and Storefront Systems Subjected to Seismic and Wind Induced Interstory Drifts*, AAMA 501.4–09 (Schaumburg, IL: American Architectural Manufacturers Association, 2009).

AAMA. *Standard Test Method for Water Penetration of Windows, Curtain Walls and Doors Using Dynamic Pressure*, AAMA 501.1–05 (Schaumburg, IL: American Architectural Manufacturers Association, 2005).

AAMA. *Test Method for Thermal Cycling of Exterior Walls*, AAMA 501.5–07 (Schaumburg, IL: American Architectural Manufacturers Association, 2007).

ACI. *Building Code Requirements for Structural Concrete*, ACI 318 (Farmington Hills, MI: American Institute of Concrete Construction, 2014).

ACI. *Specification for Tolerances for Concrete Construction and Materials*, ACI 117 (Farmington Hills, MI: American Institute of Concrete Construction, 2010).

ADA. "2010 ADA Standards for Accessible Design." Available at: www.ada.gov/regs2010/2010ADAStandards/2010ADAstandards.htm#designconstruction (accessed February 7, 2017).

AISC. *Specification for Structural Steel Buildings*, AISC 360 (Chicago, IL: American Institute of Steel Construction, 2010).

AISC. *Vibrations of Steel-Framed Structural Systems due to Human Activity*, Steel Design Guide Series 11 (Chicago, IL: American Institute of Steel Construction, 2016).

American Forest & Paper Association. *Heavy Timber Construction* (Washington, D.C.: American Forest & Paper Association, 2003).

ANSI. *Motors and Generators*, ANSI/NEMA MG 1–2011 (Washington, D.C.: American National Standards Institute, 2011).

API. *API Standard 670: Machinery Protection Systems* (Washington, D.C.: American Petroleum Institute, 2014).

ASCE. *Blast Protection of Buildings*, ASCE 59–11 (Reston, VA: American Society of Civil Engineers, 2011).

ASCE. *Minimum Design Loads and Associated Criteria for Building and Other Structures*, ASCE/SEI 7 (Reston, VA: American Society of Civil Engineers, 2016).

ASCE. *Seismic Rehabilitation of Existing Buildings*, ASCE 41–13 (Reston, VA: American Society of Civil Engineers, 2013).

ASCE. *Standard Calculation Methods for Structural Fire Protection*, ASCE/SEI/SFPE 29 (Reston, VA: American Society of Civil Engineers, 2005).

ASNT. *Personnel Qualification and Certification in Nondestructive Testing*, SNT-TC-1A (Columbus, OH: American Society for Nondestructive Testing, 2016).

ASTM. *Definitions of Terms Relating to Fatigue*, ASTM E1150 (West Conshohocken, PA: ASTM International, 1993).

ASTM. *Standard Practice for Classification of Soils for Engineering Purposes (Unified Soil Classification System)*, ASTM D2487–11 (West Conshohocken, PA: ASTM International, 2011).

ASTM. *Standard Practice for Determining Load Resistance of Glass in Buildings*, ASTM E1300–12ae1 (West Conshohocken, PA: ASTM International, 2012).

ASTM. *Standard Specification for Architectural Cast Stone*, ASTM C1364–16 (West Conshohocken, PA: ASTM International, 2016).

ASTM. *Standard Specification for Glazed Brick (Single Fired, Brick Units)*, ASTM C1405–16 (West Conshohocken, PA: ASTM International, 2016).

ASTM. *Standard Specification for Heat-Strengthened and Fully Tempered Flat Glass*, ASTM C1048–12e1 (West Conshohocken, PA: ASTM International, 2012).

ASTM. *Standard Specification for Installation of Steel Framing Members to Receive Screw-Attached Gypsum Panel Products*, ASTM C754–15 (West Conshohocken, PA: ASTM International, 2015).

ASTM. *Standard Specification for Laminated Architectural Flat Glass*, ASTM C1172–14 (West Conshohocken, PA: ASTM International: 2014).

ASTM. *Standard Specification for Nonstructural Steel Framing Members*, ASTM C645–14e1 (West Conshohocken, PA: ASTM International, 2014).

ASTM. *Standard Test Method for Determining Rate of Air Leakage through Exterior Windows, Curtain Walls, and Doors under Specified Pressure Differences across the Specimen*, ASTM E283–04 (West Conshohocken, PA: ASTM International, 2012).

ASTM. *Standard Test Methods for Liquid Limit, Plastic Limit, and Plasticity Index of Soils*, ASTM D4318–10e1 (West Conshohocken, PA: ASTM International, 2010).

ASTM. *Standard Test Method for Water Penetration of Exterior Windows, Skylights, Doors, and Curtain Walls by Uniform Static Air Pressure Difference*, ASTM E331–00 (West Conshohocken, PA: ASTM International, 2009).

ASTM. *Standard Test Methods for Fire Tests of Building Construction and Materials*, ASTM E 119 (West Conshohocken, PA: ASTM International, 2016).

Augustin, S., and Fell, D. *Wood as a Restorative Material in Healthcare Environments* (Pointe-Claire, QC: FP Innovations, 2015), 17–19.

AWC. *ASD/LFRD Manual, National Design Specifications for Wood Construction* (Leesburg, VA: American Wood Council, 2012).

AWS. *Structural Welding Code—Steel*, D1.1 (Miami, FL: American Welding Society, 2015).

Bayer, C. *AIA Guide to Building Life Cycle Assessment in Practice*, ed. R. Gentry, trans. S. Joshi (Washington, D.C.: The American Institute of Architects, 2010), 22.

Biomimicry Institute. "What is Biomimicry?" Available at: https://biomimicry.org/what-is-biomimicry/#.V55evvkrJhE (accessed February 7, 2017).

Bohnhoff, D.R. "Modeling Soil Behavior with Simple Springs, Part 1," *Frame Building News* (April 2014): 49–54.

Brand, S. *How Buildings Learn: What Happens after They're Built* (London: Penguin Books, 1995).

Brockenbrough, R.L. *AISC Design Guide 15: AISC Rehabilitation and Retrofit Guide: A Reference for Historic Shapes and Specifications* (Chicago, IL: American Institute of Steel Construction, 2002).

Bruntland Commission. *Our Common Future: Report of the World Commission on Environment and Development* (Oslo: Bruntland Commission, 1987), 16.

CCPS. *Guidelines for Vapor Cloud Explosion, Pressure Vessel Burst, BLEVE and Flash Fire Hazards* (New York: CCPS, 2010).

CESMD. "Center for Engineering Strong Motion Data." Available at: www.strongmotioncenter.org/cgi-bin/CESMD/archive.pl (accessed February 17, 2017).

Chopra, A. *Dynamics of Structures: Theory and Applications to Earthquake Engineering*, 3rd ed. (Upper Saddle River, NJ: Prentice-Hall, 2006).

Clark County Building Department. *Preparation of a Quality Systems Manual for Structural Steel Fabricators*, TG-1–06a (Clark County, NV: Clark County Building Department, 2006).

Colliery Engineer Company. *A Treatise on Architecture and Building Construction*, Vol. II, 1st ed. (Scranton, PA: Colliery Engineer Company, 1899).

Crosby, P. *Quality Is Free: The Art of Making Quality Certain* (New York: McGraw-Hill, 1979).

Das, B.M., and Nagaratnam, S. "Settlements of Shallow Foundations on Granular Soil—An Overview," *International Journal of Geotechnical Engineering* 1 (2007): 19–29.

Das, D.M. *Principles of Foundation Engineering* (Independence, KY: Brooks/Cole-Thompson Learning, 2004).

Dekay, M. *Integral Sustainable Design: Transformative Perspectives* (New York: Routledge, 2011).

Doan, A. "Biometric Architecture: Green Building in Zimbabwe Modeled after Termite Mounds," *Inhabit*, November 29, 2012, 3. Available at: http://inhabitat.com/building-modelled-on-termites-eastgate-centre-in-zimbabwe (accessed February 7, 2017).

DoD. *Unified Facilities Criteria 4–023–03 Design of Buildings to Resist Progressive Collapse* (Washington, D.C.: Department of Defense).

Eisner, H., and Manning, W. "One Meridian Plaza Fire," *Fire Engineering* 144, no. 8 (1991): 51–58.

FEMA. "Building Codes." Available at: www.fema.gov/building-codes (accessed February 20, 2017).

FEMA. *NEHRP Guidelines for the Seismic Rehabilitation of Buildings*, FEMA 273 (Washington, D.C.: Federal Emergency Management Agency, 1997).

FEMA, *NEHRP Guidelines for the Seismic Rehabilitation of Buildings*, FEMA 273 (Washington, D.C.: Federal Emergency Management Agency, 2000).

FEMA. *Next-Generation Performance-Based Seismic Design Guidelines*, FEMA-445 (Washington, D.C.: Federal Emergency Management Agency, 2006), 85.

FEMA. *Seismic Performance Assessment of Buildings*, Vol. 1, *Methodology*, FEMA P-58–1 (Washington, D.C.: Federal Emergency Management Agency, 2012).

FEMA. *Seismic Performance Assessment of Buildings*, Vol. 2, *Implementation Guide*, FEMA P-58–2 (Washington, D.C.: Federal Emergency Management Agency, 2012).

FEMA. *Seismic Performance Assessment of Buildings*, Vol. 3, *Supporting Electronic Materials and Background Documentation*, FEMA P-58–3 (Washington, D.C.: Federal Emergency Management Agency, 2012).

FEMA. *Seismic Performance Assessment of Buildings*, Vol. 4, *Methodology for Assessing Environmental Impacts*, FEMA P-58–4 (Washington, D.C.: Federal Emergency Management Agency, 2012).

FHWA. *Corrosion Costs and Preventative Strategies in the United States*, report FHWA-RD-01–156 (Washington, D.C.: Federal Highway Administration, 2002).

Fleming, R. *Design Education for a Sustainable Future* (New York: Routledge, 2011).

Fletcher, I. "Performance of Concrete in Fire: Review of the State of the Art, with a Case Study of the Windsor Tower Fire," *Proceedings of the 4th International Workshop for Structures in Fire*, Vol. 2, 2006.

"Floor Vibrations," SteelConstruction.info. Available at: www.steelconstruction.info/Floor_vibrations#Walking_paths (accessed February 20, 2017).

Gross, J., and McAllister, T. *Structural Fire Response and Probable Collapse Sequence of the World Trade Center Towers*, NIST NCSTAR 1–6 (Gaithersburg, MD: National Institute of Standards and Technology, 2005).

Hall, J.R. *U.S. Experience with Sprinklers and Other Automatic Fire Extinguishing Equipment* (Quincy, MA: National Fire Protection Association, 2009).

Hamburger, R.O. Email message to author, July 2, 2016.

Harmathy, T.Z. "Ten Rules of Fire Endurance Ratings," *Fire Technology* 1, no. 2 (May 1965): 93–102.

Hasanbas, E. "The Effects of Vibratory Screens on Supporting Structures" (M.S. thesis, University of Utah, 2013).

Hoeppner, D.W. "Industrial Significance of Fatigue Problems," in *Fatigue and Fracture*, ASM Handbook, Vol. 19 (Materials Park, OH: ASM International, 1996), 3.

Holtz, R.D., and Kovacs, W.D. *An Introduction to Geotechnical Engineering* (Englewood Cliffs, NJ: Prentice-Hall, 1981).

Hosey, L. *The Shape of Green: Aesthetics, Ecology and Design*, 2nd ed. (Washington, D.C.: Island Press, 2012), 7.

ICC. *International Building Code* (Washington, D.C.: International Code Council, 2012).

ICC. *International Building Code* (Washington, D.C.: International Code Council, 2015).

Intergovernmental Panel on Climate Change. *Climate Change and Biodiversity*, IPCC Technical Paper V (Geneva, Switzerland: Intergovernmental Panel on Climate Change, April 2002).

ISO. *Mechanical Vibration of Machines with Operating Speeds from 10 to 200 rev/s*, ISO 2372 (Vernier, Switzerland: International Standards Organization, 1974).

ISO. *Vehicle Security Barriers—Part 1: Performance Requirement, Vehicle Impact Test Method and Performance Rating*, IWA 14–1:2013 (Vernier, Switzerland: ISO, 2013).

LaMalva, K. "Thermal Response of Steel Structures to Fire: Test versus Field Conditions," *Journal of Fire Protection Engineering* 21, no. 4 (2011): 285–299.

LaMalva, K., Barnett, J., and Dusenberry, D. "Failure Analysis of the World Trade Center 5 Building", *Journal of Fire Protection Engineering* 19, no. 4 (2009): 261–274.

Liel, A.B., Haselton, C.B., and Deierlein, G.G. "How Have Changes in Building Code Provisions for Reinforced Concrete Frame Structures Improved Seismic Safety?" Available at: www.csuchico.edu/structural/documents/conference_workshop/Pub_Conf_2006_8NCEE_Liel.pdf (accessed February 17, 2017).

McAllister, T. *Structural Response and Probable Collapse Sequence of World Trade Center Building 7*, NIST NCSTAR 1–9 (Gaithersburg, MD: National Institute of Standards and Technology, 2008).

McMullin, P.W. "General Principles of Fatigue and Fracture, Part 1," *Structure Magazine*, August (Reedsburg, WI: Copper Creek, 2016).

Muir, L.S. "Rehabilitation and Retrofit of Existing Steel Structures." Available at: www.larrymuir.com/Documents/Existing Structures.pdf (accessed February 20, 2017).

The Natural Step. The Natural Step Program. Available at: www.thenaturalstep.org/ (accessed February 7, 2017).

Needham, C. E. *Blast Waves* (New York: Springer, 2010).

NFPA. *Standard for the Installation of Sprinkler Systems*, NFPA 13 (Quincy, MA: National Fire Protection Association, 2016).

NIST. "Measures of Building Resilience and Structural Robustness Project." Available at: www.nist.gov/el/building_materials/structures/mbrsr.cfm (accessed February 20, 2017).

NPS. "The Secretary of the Interior's Standards." Available at: www.nps.gov/tps/standards.htm (accessed February 20, 2017).
Peck, R.B., Hanson, W.E., and Thornburn, T.H. *Foundation Engineering* (New York: John Wiley, 1974).
Pellini, W.S. *Evolution of Engineering Principles for Fracture-Safe Design of Steel Structures*, NRL Report 6957 (Washington, D.C.: Naval Research Laboratory, 1969), 87.
Post, N. "9/11 Blazes Debunk Code Assumptions in Open-Plan Offices," *Engineering News Record* (August 2013).
Qin, X., Chouw, N., and Larkin, T. "Effect of Higher Modes on Structural Response with Nonlinear Soil-Foundation-Structure Interaction." Paper presented at ECCOMAS Congress, Crete, Greece, June 5–10, 2016.
Ricker, D.T. "Field Welding to Existing Steel Structures," *Engineering Journal*, first quarter (1988): 1–16.
Ruskin, J. *The Lamp of Memory* (London: Penguin, 2008), 233.
Salama, A.M. "Mediterranean Visual Messages: The Conundrum of Identity, Isms and Meaning in Contemporary Egyptian Architecture," *Archnet-IJAR International Journal of Architectural Research* 1, no. 1 (2007): 87.
SFPE. *Engineering Standard on Calculating Fire Exposures to Structures*, SFPE S.01 (Gaithersburg, MD: Society of Fire Protection Engineers, 2001).
Shoesmith, D.W. "Kinetics of Aqueous Corrosion," in *Corrosion*, ASM Handbook, Vol. 13 (Materials Park, OH: ASM International, 1987), 29–36.
Skidmore, Owings & Merrill. *Intelligent Densities/Vertical Communities: A New Housing Model for London*, NLA Breakfast Talk (London: NLA London's Centre for the Built Environment, July 2015), 32.
Smith, P.D., and Hetherington, J.G. *Blast and Ballistic Loading of Structures* (Boca Raton, FL: CRC Press, 1994).
SSPE. *Engineering Standard on Calculation Methods to Predict the Thermal Performance of Structural and Fire Resistive Assemblies*, SFPE S.02 (Gaithersburg, MD: Society of Fire Protection Engineers, 2015).
Starossek, U., and Haberland, M. "Disproportionate Collapse: Terminology and Procedures," *Journal of Performance of Constructed Facilities* 24, no. 6 (December 2010): 519–528.
TMS. *Building Code Requirements and Specification for Masonry Structures*, TMS 402–13 (Longmont, CO: The Masonry Society, 2013).
Trautwine, J.C. *Civil Engineers' Pocket Book* (New York: John Wiley, 1906).
UL. *Standard for Fire Tests of Building Construction and Materials*, UL 263 (Fremont, CA: Underwriters Laboratories, 2015).
UL. *UL Fire Resistance Directory*, UL Design No. N706 (Fremont, CA: Underwriters Laboratories, 2015).
University of Pennsylvania. "Penn Celebrates Completion of $24.5 Million Renovation of Historic Arts, Research and Culture House." Available at: https://news.upenn.edu/news/penn-celebrates-completion-245-million-renovation-historic-arts-research-and-culture-house (accessed February 20, 2017).

University of Pennsylvania Archives. “Christian Association.” Available at http://hdl.library.upenn.edu/1017/d/archives/20001026025 (accessed February 20, 2017).

U.S. Army. *Dewatering and Groundwater Control*, Technical Manual TM 5–818–5/AFM 88–5 (Washington, D.C.: Joint Departments of the Army and Air Force, 1983).

U.S. Army. *Soils and Geology Procedures for Foundation Design of Buildings and Other Structures*, Technical Manual TM 5–818–1/AFM 88–3, Chap. 7 (Washington, D.C.: Joint Departments of the Army and Air Force, 1983).

“Why the Building Sector” Architecture 2030. Available at: http://architecture2030.org/buildings_problem_why/ (accessed February 20, 2017).

Index